£25·00

MASONS

Solicitors
& Privy Council Agents

Health and safety law for the construction industry

Masons' guide

Susan Fink

Christopher Dering, Consulting Editor

Thomas Telford

Published by Thomas Telford Publishing, Thomas Telford Services Ltd, 1 Heron Quay, London E14 4JD

First published 1997

Distributors for Thomas Telford books are
USA: American Society of Civil Engineers, Publications Sales Department, 345 East 47th Street. New York, NY 10017 – 2398
Japan: Maruzen Co. Ltd, Book Department. 3 – 10 Nihonbashi 2-chome, Chuoku, Tokyo 103
Australia: DA Books and Journals, 648 Whitehorse Road, Mitcham 3132, Victoria

Figures 2, 3, 8 and 11 courtesy of Crown copyright. Crown copyright is repro-duced with the permission of the Controller of Her Majesty's Stationery Office.

A catalogue record for this book is available from the British Library

ISBN: 0 7277 2602 1

Coventry University

Throughout this book the personal pronouns 'he', 'his', etc. are used when referring to 'the contractor', 'the client', etc. for reasons of readability. Clearly it is quite possible these hypothetical characters may be female in 'real-life' situations, so readers should consider these pronouns to be gram-matically neuter in gender, rather than masculine, in all cases.

Typeset by Ian Kingston Editorial Services, Nottingham, UK
Printed in Great Britain by Redwood Books, Trowbridge, Wiltshire

PO 1925.
1-7-98

Preface

Health and safety legislation should be a fundamental part of any successful construction project, much the same way as are the designs, equipment and materials. Unfortunately, all too often health and safety is dismissed as 'someone else's responsibility' or too complex to bother with. That attitude is no longer acceptable to the responsible employer or to the enforcing authorities.

For these reasons it was felt that a guidebook to health and safety law in the construction industry was essential. By summarising the health and safety legislation that was in effect on 1 August 1997, the author hopes to draw your attention to the variety of duties that every construction employer, self-employed person, or site occupier may have. In addition, the author hopes to eliminate some of the complexity, as well as some of the mystery, of this legislation.

Having said that, this book is not intended to be the last word on health and safety for the construction industry — anyone familiar with the volumes of health and safety legislation currently in force will understand why. Rather, the author has tried to highlight the main health and safety duties in a convenient and readable format, and to identify any related legislation, approved codes of practice and guidance in the extensive bibliographies provided in this book. As such, while this book will not be able to answer all of your questions, it should point you in the right direction toward information that can assist you in developing your own health and safety answers.

A final word on the organisation of this book. The author has had to completely exclude certain aspects of construction safety law in the interests of simplicity. For that reason the complex rules on oil and gas safety and the rules relating to mines and quarries, for example, have not been included. Furthermore, details on the Building Regulations, as amended, have not been highlighted, with a few exceptions. While these disciplines are clearly important, the author felt them to be sufficiently specialised to require independent treatment.

The author wishes to express her thanks to HM Stationery Office for permitting her to reproduce a number of its publications in this book.

Biographies of the authors

SUSAN FINK

Susan Fink is both a solicitor and a US Attorney specialising in the areas of safety and environmental law. Before coming to the UK in 1990, Susan worked for several years in a large Chicago law firm where she practised in the areas of environmental and property law, which she used as the basis for her current practice at Masons.

Susan advises companies of all types on procedural and compliance issues with respect to health and safety and environmental law, as well as advising on general risk management issues. Further, she regularly acts for clients facing civil and/or criminal actions involving safety and environmental issues, such as oil and gas safety, asbestos licensing, nuisance, and the Construction (Design and Management) Regulations.

In addition to writing a bi-monthly column on safety developments for the *Safety and Health Practitioner*, Susan has published many articles and is a regular speaker at conferences on both environmental and safety law. Susan is an associate member of the Institute of Occupational Safety and Health.

CHRISTOPHER DERING

Chris joined Masons in 1989, after a period as a lecturer at Exeter College, Oxford, and has been a Partner, working from the firm's London office, since 1992. He practices in the fields of construction, engineering and major projects and has been responsible for the firm's health and safety group.

In terms of construction experience, he has been involved in a wide variety of contentious and non-contentious matters, including in recent times the litigation surrounding the Channel Tunnel Project. He has also advised on land development (including land subject to heavy contamination) and water concession projects in developing areas abroad.

His health and safety experience has mainly been in relation to compliance issues within the construction industry.

Chris has lectured and written on health and safety both publicly and in-house, for government organisations and others, and was active in the development of the Construction (Design and Management) Regulations.

Chris is a former editor of the *Jersey Law Reports*, a contributor to *Service Level Agreements*, co-editor and co-author of *Eco-Management and Eco-Auditing: Environmental Issues in Business*, and is a member of the Editorial Board of *Facilities Management Legal Update*.

Contents

Chapter 1 Introduction to health and safety law   1

1.1. Introduction   1

1.2. Health and Safety Statutory Law   2
    1.2.1. The Health and Safety at Work etc. Act 1974   2
    1.2.2. Statutory instruments, ACoP and guidance   2
    1.2.3. 'So far as is reasonably practicable'   4

1.3. The Management Regulations   4
    1.3.1. Assessment   5
    1.3.2. Appoint a competent person   5
    1.3.3. Co-operate and co-ordinate with other employers 6
    1.3.4. Provide information and training   6

1.4. Statutory Enforcement Actions   6
    1.4.1. Improvement and Prohibition Notices   7
    1.4.2. Criminal prosecutions   8
    1.4.3. Directors' and Officers' liability   8
    1.4.4. Directors' disqualification   9
    1.4.5. Burden of proof   9
    1.4.6. Manslaughter actions   9

1.5. Civil Liability for Health and Safety   10
    1.5.1. Breach of Contract   10
    1.5.2. Negligence   11
    1.5.3. Breach of Statutory Duty   11

1.6. Employers' Liability Insurance   12

Chapter 2 Health and safety liabilities   13

2.1. Introduction   13

2.2. Employers' Liability for Safety of Employees   14

2.3. Employers' Liability for the Safety of
Non-employees — Section 3 of the HSW Act   15
    2.3.1. R v. *Associated Octel*   16

2.4. Occupiers' Liability for the Safety of Non-employees   18
    2.4.1. Section 4 of the HSW Act   18
    2.4.2. Occupiers' Liability Acts   18

2.5. The Health and Safety Policy   19
    2.5.1. HSW Act policy requirements   19
    2.5.2. Health and safety information for employees   20

2.6. Other Relevant Legislation   21

Chapter 3 The Construction (Design and Management) Regulations   23

3.1. Historical Development of Regulations   23
    3.1.1. The Temporary or Mobile Construction Sites
Directive   23

3.2. The Construction (Design and Management)
Regulations 1994   24
    3.2.1. What makes these Regulations special   24
    3.2.2. The Health and Safety Plan   25
    3.2.3 The Health and Safety File   27
    3.2.4. The Client   27
    3.2.5. The Designer   30
    3.2.6. The Planning Supervisor   32
    3.2.7. The Principal Contractor   33
    3.2.8. The Contractor   36

3.3. Exclusion of Civil Liability   37

CONTENTS

Chapter 4   Equipment safety   39

4.1. Introduction   39

4.2. Equipment Safety — Generally   39
      4.2.1. The Provision and Use of Work Equipment
              Regulations 1992   39
      4.2.2. Proposed Work Equipment Regulations   42
      4.2.3. Supply of Machinery (Safety) Regulations   42

4.3. Lifting Equipment   44
      4.3.1. Construction (Lifting Operations)
              Regulations 1961   45
      4.3.2. Cranes   46
      4.3.3. Hoists   51

4.4. Access Equipment   52
      4.4.1. Statutory requirements   53
      4.4.2. Access equipment guidance   53

4.5. Lifts Regulations   54

Chapter 5   Workplace safety   55

5.1. Introduction   55

5.2. The Construction (Health, Safety and Welfare)
      Regulations 1996   56
      5.2.1.   Application of the Regulations   56
      5.2.2.   Safe place of work   57
      5.2.3.   Fall prevention   58
      5.2.4.   Falling objects   58
      5.2.5.   Stability of structures   59
      5.2.6.   Demolition or dismantling   59
      5.2.7.   Explosives   59
      5.2.8.   Excavations   59
      5.2.9.   Vehicles and traffic routes   60
      5.2.10. Emergency procedures   60

5.2.11. Welfare   61
5.2.12. Training   61
5.2.13. Inspection   61
5.2.14. Miscellaneous requirements   62

5.3. The Workplace (Health, Safety and Welfare)
      Regulations 1992   62
      5.3.1. Maintenance and cleanliness of workplace   62
      5.3.2. Ventilation, temperature and lighting   63
      5.3.3. Space and workstation requirements   63
      5.3.4. Fall prevention   63
      5.3.5. Floors and traffic routes   64
      5.3.6. Windows, doors, gates, walls and skylights   64
      5.3.7. Other welfare facilities   64

5.4. Other Workplace Standards   65
      5.4.1. Lighting   65
      5.4.2. Workplace noise   66

Chapter 6   Fire safety   69

6.1. Introduction   69

6.2. Fire Certification   69
      6.2.1. The Fire Precautions Act 1971   70
      6.2.2. Contents of fire certificate   70

6.3. Safety Equipment/Fire Precautions for Exempt Premises 71

6.4. Fire Precautions (Workplace) Regulations 1977   72

6.5. Building Regulations   73

6.6. Special Premises   74

6.7. Other Fire Legislation   76

Chapter 7   Personal protective equipment   79

7.1. Introduction   79

7.2. The Personal Protective Equipment at Work
Regulations 1992   81
    7.2.1. Assess need for PPE   81
    7.2.2. Provide effective PPE   81
    7.2.3. Maintain and repair PPE   83
    7.2.4. Provide PPE accommodation   83
    7.2.5. Provide information, instruction and training 84
    7.2.6. Ensure use of equipment   84
    7.2.7. Employees' duties   85

7.3. Other Personal Protective Equipment Legislation   85
    7.3.1. The Construction (Head Protection)
            Regulations 1989   86
    7.3.2. The Noise at Work Regulations 1989   87
    7.3.3. The Control of Asbestos at Work
            Regulations 1987   87
    7.3.4. The Control of Substances Hazardous to
            Health Regulations 1994   88

7.4. Guidance on Personal Protective Equipment   88

Chapter 8   Hazardous substances   89

8.1. Introduction   89

8.2. Control of Substances Hazardous to Health Regulations 89
    8.2.1. Definition   90
    8.2.2. Employers' duties under COSHH   90
    8.2.3. MEL and OES   91

8.3. Asbestos   92
    8.3.1. The Control of Asbestos at Work
            Regulations 1987   92
    8.3.2. The Control of Asbestos (Amendment)
            Regulations 1992   93
    8.3.3. The Asbestos (Licensing) Regulations 1983   94
    8.3.4. The Asbestos (Prohibitions) Regulations 1992 94

8.4. Ionising Radiation   94
    8.4.1.   The Ionising Radiations Regulations 1985   94
    8.4.2.   Ionising Radiations (Outside Workers)
            Regulations 1993   96

8.5. Explosives   97

8.6. Other relevant statutes   97

Chapter 9 Accident reporting and investigation   99

9.1. Introduction   99

9.2. The RIDDOR Regulations   99
    9.2.1.   Definitions   100
    9.2.2.   Reporting requirements   101
    9.2.3.   Record-keeping requirements   102

9.3. Investigating Accidents   107
    9.3.1.   Executive's power to investigate   107
    9.3.2.   Safety representatives' power to investigate 109

Chapter 10 First-aid on the site   111

10.1. Introduction   111

10.2. The First-aid Regulations   111
    10.2.1. Duty to provide first-aid equipment   111
    10.2.2. Assessing first-aid requirements   112
    10.2.3. Duty to provide first-aiders and/or
           appointed persons   113
    10.2.4. Duty to inform   114
    10.2.5. Duty applies to the self-employed   114
    10.2.6. Record keeping   114

10.3. First-aid boxes and rooms   115

10.4. Exclusions   116

CONTENTS

Chapter 11 Miscellaneous health and safety issues   **119**

11.1. Introduction   119

11.2. Manual handling   119
    11.2.1. Duty to limit manual handling   120
    11.2.2. Duty to assess manual handling tasks
         remaining   120
    11.2.3. Employees' duty   122
    11.2.4. Guidance on Manual Handling   122

11.3. Display screen equipment   125
    11.3.1. Definitions   125
    11.3.2. Duty to assess workstations   126
    11.3.3. Eye and eyesight examinations   128
    11.3.4. Information and training   129

11.4. Safety signs/signals   129
    11.4.1. Safety signs requirements   129
    11.4.2. Other duties   131

Table of statutes   **133**

Table of statutory instruments and orders   **135**

Bibliography   **139**

Index   **143**

1

Introduction to
health and safety law

1.1. INTRODUCTION

Before embarking on an explanation of the details of health and safety duties for the construction industry, it is essential to have an understanding as to the source of those duties. Accordingly, this Chapter will outline the two separate regimes that make up health and safety law in the UK, namely statutory law and civil law.

In addition, this Chapter will outline the legal consequences that may follow on from a breach of either of those laws. In particular, if a person breaches a health and safety *statute* he will have committed an offence which is punishable by some type of enforcement proceeding. Those enforcement proceedings are described in detail in Section 1.4.

If, on the other hand, a person breaches a health and safety civil law, the injured party can bring a civil action for damages, alleging, for example, negligence, breach of contract, or the tort of breach of statutory duty. Those civil actions are described in detail in Section 1.5.

1.2. HEALTH AND SAFETY STATUTORY LAW

Health and safety statutes first came into being over two hundred years ago. Those early statutes were typically comprised of a haphazard set of rules specifying or prohibiting certain actions. As such, they tended to be prescriptive and inflexible, leaving no scope for individual differences amongst businesses. That statutory regime continued virtually unabated until the 1960s, with the creation of statutes like the Factories Act 1961 and the Offices, Shops and Railway Premises Act 1963.

Despite the existence of such laws, accident rates continued to rise. This fact led the Government to commission the first comprehensive review of health and safety law in the UK, which was embodied in 'The Robens Report'. That Report suggested that health and safety law should move away from a system based on detailed and prescriptive rules to a system based on principles of good practice, thus allowing companies to set their own safety standards and to develop their own safety procedures.

1.2.1. The Health and Safety at Work etc. Act 1974

The Robens Report led to the adoption of the Health and Safety at Work etc. Act 1974 (the HSW Act) and a new era in health and safety legislation which was based on the principles of risk assessment and goal-setting, rather than compliance with inflexible and prescriptive rules.

The HSW Act is the focus of all health and safety statutory law in the UK. Specifically, in ss. 2 – 9, the Act outlines all of the general principles which underlie all other health and safety legislation. Table 1 provides a list of those main duties.

In view of their fundamental importance, Chapter 2 has been devoted to a discussion of these principles.

1.2.2. Statutory instruments, ACoP and guidance

As Table 1 illustrates, the HSW Act provides little in the way of detail or standards for employers to follow. For this reason, the Act

2

Table 1. The Health and Safety at Work etc. Act 1974

HSW Act Section	Duties created
Section 2	Imposes duties on employers
	• to ensure, so far as is reasonably practicable, the health and safety and welfare of all employees
	• to ensure the provision and maintenance of safe plant and systems of work for all employees, so far as is reasonably practicable
	• to provide safe systems with regard to the use, storage and transport of articles and substances for all employees, so far as is reasonably practicable
	• to provide such information, instruction, training and supervision as is necessary to ensure the health and safety at work of employees, so far as is reasonably practicable
	• to maintain a safe place of work and to provide and maintain safe access to and egress from that place of work for all employees, so far as is reasonably practicable
	• to provide and maintain a working environment for employees that is safe without risk to health, so far as is reasonably practicable, and
	• to provide and maintain a written statement of safety detailing the safety policy of that company and the arrangements for implementing it.
Section 3	Imposes duty on employers and the self-employed to ensure the safety of persons other than their own employees.
Section 4	Imposes duty on persons in control of or concerned with premises to ensure the safety of persons on those premises.
Section 5	Imposes duty on persons in control of premises to limit or prevent the release of harmful emissions into the atmosphere.
Section 6	Imposes duty on designers and manufacturers of articles and substances to ensure that they are safe during use.
Section 9	Prohibits employers from charging their employees for safety equipment or systems provided.

empowers the Health and Safety Commission (the HSC) to propose statutory instruments to supplement those general duties. As a result, the HSW Act is now supplemented and supported by a vast range of statutory instruments, regulations and orders.

In addition, the HSC has developed a number of Approved Codes of Practice (ACoP) and/or Guidance Notes to provide interpretative assistance and practical guidance to employers when attempting to comply with the legislation. While failure to comply with the provisions of an ACoP is not an offence, the failure itself can be used as direct evidence in a prosecution of a breach of the underlying legislation. No legal significance has been given to a breach of a Guidance Note, however.

1.2.3. 'So far as is reasonably practicable'

It is important to note that, virtually without exception, the duties imposed by the HSW Act and the subsidiary legislation passed there-under are limited by the phrase 'so far as is reasonably practicable'. That phrase has the effect of permitting an employer, or any duty-holder, to conduct a cost/benefit analysis in respect of his health and safety systems. In other words, it allows the duty-holder to calculate whether the benefits afforded by the addition of certain health and safety procedures are outweighed by the costs (in terms of time, in-convenience, money, etc.) of those additional procedures. If so, those precautions need not, in theory, be taken.

1.3. THE MANAGEMENT REGULATIONS

Clearly one, if not *the* most important health and safety statutory in-struments to come into force since the HSW Act, is the Management of Health and Safety at Work Regulations 1992. Those Regulations were introduced to stem the tide of rising accidents in the 1980s, which was thought to be a function of the fact that the HSW Act failed to explicitly require employers to develop effective safety man-agement systems.

4

The Management of Health and Safety at Work Regulations 1992, as amended by the Management of Health and Safety At Work (Amendment) Regulations 1994 (collectively, the Management Regulations), therefore impose a number of management duties on employers and others for virtually every workplace—including construction sites. Given the fundamental effect of these Regulations, it is necessary to summarise the main duties contained therein.

1.3.1. Assessment

Regulation 2 of the Management Regulations is clearly one of the most important obligations created by those Regulations. In particular, Regulation 2 obliges all employers to perform a risk assessment of his operations.

To do this, the employer must first identify

- the hazards present in his operations (which hazards may go beyond the boundaries of his workplace)
- the type and severity of any injury that could result from those hazards, and
- the likelihood that such an injury could occur.

Thereafter, the employer must develop appropriate procedures to eliminate the hazards found or, if elimination is not possible, to reduce the risk of injury, so far as is reasonably practicable. The ACoP prepared for the Regulations provides employers with advice on the best methods for reducing risks.

1.3.2. Appoint a competent person

Every employer must appoint one or more competent persons to advise and assist him in his statutory health and safety duties. That 'competent person' may be an employee or a consultant, provided that he has sufficient training, knowledge and experience to perform the task.

1.3.3. Co-operate and co-ordinate with other employers

In addition to the duty to prepare risk assessments and to develop safety procedures, Regulation 9 of the Management Regulations is of particular relevance to employers in the construction industry. Regulation 9 requires employers and self-employed persons who share a worksite to

- co-operate with one another in respect of health and safety, and
- co-ordinate their health and safety procedures and information.

Clearly these duties are of relevance on a construction site, where you may have several employers and self-employed persons working together in close proximity. It is not surprising, therefore, that these duties of co-operation and co-ordination are echoed in the management systems prescribed in the Construction (Design and Management) Regulations 1994, which are described in detail in Chapter 3.

1.3.4. Provide information and training

Employers must provide their employees with suitable and sufficient information and training on

- the risks identified in their assessment
- the emergency procedures developed to reduce those risks, and
- the identity of the competent person.

This training and information must be monitored and reviewed, and updated as appropriate.

Table 2 contains a complete list of the main duties imposed by the Management Regulations.

1.4. STATUTORY ENFORCEMENT ACTIONS

Section 33 of the HSW Act makes it an offence for an employer to fail to discharge a statutory duty prescribed by that Act or by any subsidiary legislation. The form and severity of that offence will depend on

6

Table 2. The Management of Health and Safety at Work Regulations 1992 (as amended)

Regulation	Employers' duties
Regulation 3	To make suitable and sufficient risk assessments of their operations.
Regulation 4	To develop preventative and protective measures, as appropriate, and to record those measures when employing five (5) or more workers.
Regulation 5	To provide employees with health surveillance, as appropriate.
Regulation 6	To appoint a competent person to assist in implementing health and safety procedures.
Regulation 7	To establish procedures in the event of serious and imminent danger.
Regulation 8	To provide their employees with information on health and safety risks, the protective measures instituted and the names of competent persons appointed.
Regulation 9	To the extent a workplace is shared, employers must co-operate with one another in respect of health and safety and co-ordinate their health and safety measures.
Regulation 10	To the extent that another employer's employees are working on a site, provide safety information to the employer of said employees.
Regulation 11	To consider an employee's capabilities before allocating tasks.
Regulation 13	To provide temporary employees and their employers with information on the skills and surveillance associated with their positions.

the form and severity of that breach. As such, an employer may find himself faced with either an improvement or prohibition notice, or a criminal prosecution for violating health and safety laws.

1.4.1. Improvement and Prohibition Notices

The most common form of enforcement action is the issuance of an improvement or prohibition notice. An improvement notice can be served by an inspector if he believes there has been or may be in the future a breach of relevant legislation. The notice will require certain steps to be taken to correct that breach within a stated period, which must be at least 21 days.

On the other hand a prohibition notice can be served by an inspector whenever he believes that a hazardous state of affairs exists which presents a real risk of serious personal injury—whether or not there has been a breach of any specific legislation. The effect of such a notice is to order the relevant activity to be stopped—either immediately or after a stated period—until certain specified steps have been taken. An employer may appeal a prohibition or improvement notice to an industrial tribunal.

1.4.2. Criminal prosecutions

For more serious breaches of health and safety laws, a criminal proceeding may be brought before the Magistrates' Courts or the Crown Court. While most offences are triable in either court, the less serious offences can only be dealt with in the Magistrates' Court, with the more serious breaches dealt with by the Crown Court. For example, prosecutions for breaching the s. 2 duties are 'triable either way'—with the more serious offences referred to the Crown Court, whereas a prosecution for failing to comply with a prohibition or improvement notice can be heard only by the Magistrates' Court.

The maximum fine that can be imposed in the Magistrates' Court for breaches of the principal parts of the HSW Act (ss. 2 – 6) is £20 000. The maximum fine that the Magistrates can impose for other breaches of the Act or of other statutes is £5000. The Magistrates can impose custodial sentences of up to six months for breaches of certain improvement notices. On the other hand, the Crown Court can order unlimited fines and prison sentences of up to two years for certain offences, such as contravening the terms of a licence issued by the Health and Safety Executive (the HSE).

1.4.3. Directors' and Officers' liability

In the same way that individuals and companies can be prosecuted for health and safety offences, s. 37 of the HSW Act makes it clear that

8

senior managers of a company can be prosecuted for safety offences. Section 37 provides that

> Where an offence under any of the relevant statutory provisions committed by a body corporate is proved to have been committed with the consent or connivance of, or to have been attributable to any neglect on the part of any director, manager, secretary or a similar officer of the body corporate or as a person who is purporting to act in any such capacity, he, as well as the body corporate, shall be guilty of that offence and shall be liable to be proceeded against and punished accordingly.

1.4.4. Directors' disqualification

A director found guilty under s. 37 of the HSW Act may also be disqualified for up to two years from serving as a director of a company, by virtue of the Companies Directors Disqualification Act 1986. Section 2(1) of that Act provides that by having committed an indictable offence connected with the management of a company, a director can be disqualified.

1.4.5. Burden of proof

For a defendant to be convicted of a criminal offence, the prosecution must show *beyond a reasonable doubt* that the accused committed that offence. However, the HSW Act makes the prosecution's job easier by transferring some of the burden of proof to the accused in health and safety cases.

Specifically, s. 40 the HSW Act provides that the accused must show that his actions were either: (a) practicable, (b) reasonably practicable or (c) that there was no better practical means than those used to satisfy the particular duty or requirement. If the accused cannot satisfy this test, the case will be considered proven against him. (See Section 1.2.3. for a discussion of the meaning of the phrase 'reasonably practicable'.)

1.4.6. Manslaughter actions

To the extent that a breach of a health and safety statute causes a fatality it is possible that the defendant may be prosecuted under general criminal law for the offence of manslaughter. Manslaughter is defined as death caused by the unlawful act or gross criminal negligence of a person. Manslaughter, as compared to murder, is punishable by a fine and as such is applicable to both individual or company defendants.

Traditionally, prosecutions of companies for manslaughter have failed for proof problems. However, a proposal has been made by the Law Commission which would effectively eliminate those problems, and as such make it much easier to prosecute companies. Those proposals are contained in the Law Commission Paper 127, entitled *Legislating the Criminal Code: Involuntary Manslaughter*. At the time of writing it is not clear if, or when, these proposals may come into force.

1.5. CIVIL LIABILITY FOR HEALTH AND SAFETY

There are cases going back hundreds of years in respect of civil actions against defendants (typically employers) for breaching their common law safety responsibilities, usually following an accident that causes injury.

For the most part, the plaintiffs (typically employees) claim that the defendants breached their contractual duty or committed negligence. More recently, however, employees have also been claiming for damages for 'breach of statutory duty', which is a special type of civil duty. The following is a brief description of those actions.

1.5.1. Breach of Contract

The duty to take reasonable care for the safety of an employee is one of the terms of the employment contract between an employer and an employee. Sometimes this duty will be expressly stated in the contract, if not it will be implied as a matter of law.

In either case, if an employee is injured as a result of his employer's failure to take reasonable care, that employee can pursue a civil action against his employer for breach of contract.

To succeed in an action for breach of contract, the employee would have to show that

- there was an employment contract with the defendant
- the defendant breached that contract by failing to take reasonable care, and
- as result of that breach, the employee was injured.

This action can be served in conjunction with, or as an alternative to, a claim for negligence or breach of statutory duty.

1.5.2. Negligence

In addition, employers owe a common law duty of care to ensure the health and safety of their employees by providing, among other things

- safe and adequate plant and equipment
- safe premises and/or place of work
- competent and safe fellow workers, and
- a safe system of work.

If an employer fails to take *reasonable* care in these matters and an employee is injured as a result, that employee may be entitled to recover damages for negligence, provided that the employee can show that

- the employer owed him a duty of care
- the employer breached that duty of care, and
- the employee was injured as a result of that breach.

The duty of care at common law only obliges the employer to take *reasonable steps* to prevent harm. As such, an employer is entitled to balance the utility of the task with the degree of risk and he will not be held in breach of his duty of care where the foreseeable risk of injury is very slight and there was no other reasonable method of performing the same task.

1.5.3. Breach of Statutory Duty

The most recent weapon in plaintiffs' arsenal of civil claims is the tort of 'breach of statutory duty'. This action allows an injured party (typically an employee, but not necessarily) to bring an action against any person who has a statutory duty to provide for that person's health and safety. In order to succeed in a claim for breach of statutory duty, the injured party must show that

- a duty was breached that can give rise to a civil liability
- the statutory duty is owed to the plaintiff by the defendant
- there is a breach of the statutory duty by the defendant
- the damage suffered by the plaintiff is of a type which the statute was designed to prevent, and
- the injury was caused by the defendant's breach.

It is generally not possible to bring an action for breach of a statutory duty that relates solely to employee *welfare*. However, it is generally possible to bring such an action for breach of a duty that relates to employee *safety*. Typically statutory instruments (such as the Management Regulations) will state specifically if they preclude actions for breach of statutory duty.

1.6. EMPLOYERS' LIABILITY INSURANCE

Every employer is obliged to maintain employers' liability insurance by virtue of the Employers' Liability (Compulsory Insurance) Act 1969. That insurance is intended to ensure payment in the event that an employer is found liable for bodily injury or disease sustained by his employees, and arising out of and in the course of their employment. Employers are obliged to maintain coverage of at least £2 million/occurrence.

This insurance can be used to pay both the damages assessed and the costs incurred in a civil action against the employer. It can also often be used to pay the employer's costs in defending a criminal proceeding. However, it can *not* be used to pay any fines or penalties that may be imposed in a criminal proceeding, as criminal penalties cannot be insured against.

2

Health and safety liabilities

2.1. INTRODUCTION

The primary focus for all health and safety legislation is the Health
and Safety at Work etc. Act 1974 (the HSW Act) and the subsidiary
legislation passed thereunder. While there are other health and safety
Acts still in force which could affect safety on a construction site (e.g.
the Factories Act 1961 and the Offices, Shops and Railway Premises
Act 1963), they are slowly being phased out and replaced by subsidiary legislation. (See Section 2.6. for a discussion of other relevant
legislation.)

As this Chapter will demonstrate, the HSW Act imposes a wide
variety of duties on

- employers to ensure the safety of their employees (primarily s. 2 of
  the Act)
- employers to ensure the safety of persons other than their
  employees, who may be affected by their operations (s. 3 of the
  Act), and
- other persons, insofar as their activities or premises affect safety
  (s. 4 of the Act).

Other sections of the HSW Act also impose duties on employers. For a full list of those duties, see Table 3.

Table 3. Other duties under the Health and Safety at Work etc. Act 1974

HSW Act Section	Duties created
Section 3	Imposes duty on employers and the self-employed to ensure the safety of persons other than their own employees.
Section 4	Imposes duty on persons in control of or concerned with premises to ensure the safety of persons on those premises.
Section 5	Imposes duty on persons in control of premises to limit or prevent the release of harmful emissions into the atmosphere.
Section 6	Imposes duty on designers and manufacturers of articles and substances to ensure that they are safe during use.
Section 9	Prohibits employers from charging their employees for safety equipment or systems provided

2.2. EMPLOYERS' LIABILITY FOR SAFETY OF EMPLOYEES

In order to have a full appreciation of an employer's duties to ensure the safety of his employees, one must begin by looking at the general principles laid down in s. 2 of the HSW Act, which are

- to ensure, so far as is reasonably practicable, the health and safety and welfare of all employees
- to ensure the provision and maintenance of safe plant and systems of work for all employees, so far as is reasonably practicable
- to provide safe systems with regard to the use, storage and transport of articles and substances for all employees, so far as is reasonably practicable
- to provide such information, instruction, training and supervision as is necessary to ensure the health and safety at work of employees, so far as is reasonably practicable

14

- to maintain a safe place of work and to provide and maintain safe access to and egress from that place of work for all employees, so far as is reasonably practicable
- to provide and maintain a working environment for employees that is safe without risk to health, so far as is reasonably practicable, and
- to prepare a written statement detailing the safety policy of that company and the arrangements for implementing it when a company has five or more employees.

Together these serve as the basis for most other health and safety duties that are imposed on employers in respect of their employees.

2.3. EMPLOYERS' LIABILITY FOR THE SAFETY OF NON-EMPLOYEES — SECTION 3 OF THE HSW ACT

We focus in this part on the general duties of an employer to ensure the health and safety of persons *not* in his employ, under s. 3 of the HSW Act. As we will see, that section has particular relevance to the construction industry, in view of the industry's reliance on subcontracted labour and the likelihood of interaction with the general public.

Section 3(1) provides that

> It shall be the duty of every employer to conduct his undertaking in such a way as to ensure, so far as is reasonably practicable, that persons not in his employment who may be affected thereby are not thereby exposed to risks to their health or safety.

This section has the effect of imposing a duty on an employer to ensure, so far as is reasonably practicable, the health and safety of persons who may not be in his employ but who may be affected by his operations, such as the employees of independent contractors, members of the general public, members of the emergency services, etc.

15

The scope of the s. 3 duties has been the subject of a recent line of cases, beginning with the High Court's decision in *RMC Roadstone Ltd* v. *Jester* (1994). That case involved the prosecution of a company for failing to ensure the safety of two independent contractors whom it had employed to replace asbestos sheeting to its roof. During that operation one of the contractors fell through a skylight and was killed. The company who had employed the services of that contractor was convicted of breaching s. 3(1) of the Act for failing to ensure the safety of persons not in its employ.

On appeal, the Appeal Court held that the conviction under s. 3(1) could not be upheld unless it could be shown that the employer had either exercised *actual control* over the activities of the contractors or if it had a *legal duty* to exercise such control. In the case before it, the Appeal Court found that the employer had neither actual control over the contractors' activities nor any legal duty to exercise such control and reversed the conviction.

For this reason *RMC Roadstone Ltd* was seen as standing for the proposition that if an employer wished to protect himself from liability under s. 3(1) he should neither instruct nor advise contractors in their method of working or their safety procedures. However, if an employer did involve himself (whether voluntarily or otherwise) in a contractor's activities, and in doing so exercised *control* over the contractor, the contractor's activities would fall within the employer's 'conduct of undertaking' — making him eligible for prosecution under s. 3(1). As such, whether or not an employer had *control* over the activities of the contractor was relevant to determining whether there was a cause of action under s. 3(1).

2.3.1. *R v. Associated Octel*

RMC Roadstone represented the state of the law in this area until August 1994 when the Court of Appeal (Criminal Division) handed down its decision in *R* v. *Associated Octel Co. Ltd* (1994). In *Octel*, the employer operated a chemical plant and employed a firm of specialist contractors to perform maintenance and repair during the plant's annual shut-down. During that operation, a light bulb used by one of the contractor's employees burst, igniting the cleaning fluid being used and badly burning him. Both the contractor and Octel were prosecuted, with Octel convicted for failing to secure the safety of persons not in its employ under s. 3(1).

Octel appealed its conviction to the Court of Appeal, claiming that the contractor's activities were not within 'the conduct of its undertaking' because the plant was shut down at the time of the accident and because the contractor had virtually complete control of safety on site. (Having said that, it is important to the case that the contractor was following Octel's permit to work system at the time of the accident and that Octel had supplied some of the safety equipment used by the contractor's employees.)

The Court of Appeal rejected Octel's argument and denied the appeal. Octel sought leave to appeal the conviction to the House of Lords. On 14 November 1996, the House of Lords rejected all of Octel's arguments, and upheld the conviction (see *R* v. *Associated Octel Co. Ltd*, [1996] 1 WLR 1543).

In the majority opinion, Lord Hoffman considered Octel's argument that it had no case to answer in an action under s. 3(1) because the activities of the subcontractor's employee were not within the 'conduct of Octel's undertaking'. He held that normally this was a question of fact for the jury, but that he was of no doubt that a jury would find that Octel was conducting its undertaking by instructing the specialist contractor because

> Octel's undertaking was running a chemical plant at Ellesmere Port. Anything which constituted running the plant was part of the conduct of its undertaking.... [As such] it is part of the conduct of the undertaking, not merely to clean the factory, but to have the factory cleaned by contractors.

To the extent that s. 3 applies, Lord Hoffman held that an employer was obliged to take all

> reasonably practicable steps to avoid risk to the contractors' servants which arise, not merely from the physical state of the premises... but also from the inadequacy of the arrangements which the employer makes with the contractor for how they do the work.

17

To summarise, the Lords have confirmed that on virtually every occasion when an employer engages a contractor in the furtherance of his business, that employer will be subject to the duties imposed by s. 3 of the HSW Act. For this reason, it is fair to say that since the decision in *Octel* the scope of s. 3 is almost as wide as that of s. 2 of the HSW Act.

2.4. OCCUPIERS' LIABILITY FOR THE SAFETY OF NON-EMPLOYEES

2.4.1. Section 4 of the HSW Act

Section 4 of the HSW Act imposes safety responsibilities on persons other than employers.

> It shall be the duty of each person who has... control of premises... to take such measures as it is reasonable for a person in his position to take to ensure, so far as is reasonably practicable, that the premises... is or are safe and without risks to health

In this way s. 4 goes beyond the duties created by ss. 2 and 3 of the HSW Act by imposing duties on persons in 'control of premises' — whether or not they are employers. The phrase 'persons in control of premises' has been interpreted to mean the same thing as 'occupiers' at common law, i.e. any person with a sufficient degree of control over the state of the premises or over the activities of individuals thereon, even if that control is shared with other individuals.

2.4.2. Occupiers' Liability Acts

In addition to the duties imposed on occupiers by s. 4 of the HSW Act, the Occupiers' Liability Act 1957 and the Occupiers' Liability Act 1984 also place duties on occupiers to ensure the safety of individuals on their property.

In particular, the 1957 Act provides that occupiers have a duty to take reasonable care to ensure the safety of all *visitors* to the property during the term of his/her visit. The 1984 Act imposes a similar duty on occupiers, but extends its application to include *non-visitors* (i.e. trespassers), if their presence on the property is known or reasonably likely and there is a chance that an injury could occur if no action is taken. Failure of an occupier to fulfil either of these duties is violation of these *civil* statutes and a possible basis for an action for damages.

2.5. THE HEALTH AND SAFETY POLICY

2.5.1. HSW Act policy requirements

Another of the more general duties imposed on employers by the HSW Act is to prepare a health and safety policy if they employ more than five workers at any one time.

Specifically, s. 2(3) of the HSW Act provides that

> It shall be the duty of every employer to prepare and as often as may be appropriate, revise, a written statement of his general policy with respect to the health and safety at work of his employees and the organisation and arrangements for the time being in force for carrying out that policy, and to bring that statement and any revision of it to the notice of all his employees.

As health and safety policies are often quite general in tone, they are frequently supplemented by written procedures addressing specific health and safety risks. While this system is not mandatory, it is a useful format for both employers and employees.

This arrangement is particularly useful for referring to information generated by the risk assessments performed by employers under the Management of Health and Safety at Work Regulations 1992. In particular, those Regulations require employers to record all significant hazards identified in their risk assessments, as well as any procedures developed to reduce the risks of injury therefrom. (See Chapter 1 for a complete description of those Regulations.)

Employers must regularly review and update their policies and procedures. The purpose of this review is to ensure, among other things, the continuing effectiveness of the employer's policy and to ensure that procedures are still appropriate. Furthermore, the policy (and the procedures) must be given and explained to all employees every time it is updated.

2.5.2. Health and safety information for employees

Related to the duty to provide a health and safety policy are the requirements of the Health and Safety Information for Employees Regulations 1989. Those Regulations provide, in relevant part, that

> an employer shall ensure (either) that the approved poster is kept displayed in a readable condition at a place which is reasonably accessible to the employee while he is at work, and in such a position in that place as to be easily seen and read by that employee; or (he shall) give the employee the approved leaflet

These Regulations were developed to ensure that employees have access at all times to essential information on their health and safety rights and responsibilities. This is accomplished, in part, by ensuring that the approved poster and the approved leaflet contain information on

- the general health and safety duties of employers
- the general health and safety duties of employees, and
- procedural information for employees wishing to report health and safety problems, which information must include the address of the local enforcing authority and the address of the local Employment Medical Advisory Service (which Service is provided for in the HSW Act).

Please note that these Regulations were amended recently by the Health and Safety Information for Employees (Modifications and Repeals) Regulations 1995, which permit certain classes of employers to use posters or leaflets other than the approved poster or leaflet provided that they apply for and receive approval from the HSE. Employers are advised to refer to these Regulations for details.

20

2.6. OTHER RELEVANT LEGISLATION

It must be remembered that the HSW Act is not the only health and safety statute that affects construction sites. Both the Factories Act 1961 and the Offices, Shops and Railway Premises Act 1963 may apply — the former with respect to the construction site itself and the latter with respect to any part of the site used as offices.

Having said that, the importance of both of these Acts was largely depleted on 1 January 1997 when the Management of Health and Safety at Work Regulations 1992, the Provision and Use of Work Equipment Regulations 1992 and, for sites where construction work is not in progress, the Workplace (Health, Safety and Welfare) Regulations 1992 all came into full force and effect. (See Chapters 1, 4 and 5 respectively for a detailed discussion of those Regulations.)

3

The Construction (Design and Management) Regulations

3.1. HISTORICAL DEVELOPMENT OF REGULATIONS

It is fair to say that the Construction (Design and Management) Regulations 1994 (the CDM Regulations) represent one of the most fundamental revisions to construction safety law since the 1960s. In view of this, it is essential to consider these Regulations in some detail — beginning with a brief outline of their historical background.

3.1.1. The Temporary or Mobile Construction Sites Directive

The CDM Regulations were developed to implement the Temporary or Mobile Construction Sites Directive, which was adopted by the European Council in 1992. That Directive had been drafted in response to research which indicated that poor management of a construction project and, in particular, poor training, communication and planning, were directly related to levels of safety on a construction site.

That research also demonstrated a direct correlation between the design of a structure and the risk of injury to the persons involved in its construction. Furthermore, inadequate co-ordination of contractors, particularly where various activities were ongoing simultaneously or in succession, was found to be the cause of many accidents on construction sites.

3.2. THE CONSTRUCTION (DESIGN AND MANAGEMENT) REGULATIONS 1994

3.2.1. What makes these Regulations special

The CDM Regulations reflect a fundamental shift in the Health and Safety Commission's policy for construction safety in the UK, moving it away from the traditional system whereby each employer is obliged to determine the proper course of conduct for his own employees only, towards a system that obliges every employer on a construction site to become part of a health and safety management system set up for the project as a whole.

The CDM Regulations are also different from earlier health and safety legislation because they actually impose a management structure for construction sites. Imposition of that structure was deemed essential to reinforce the fact that

- clients are in a position to exercise control over the level of safety adopted by the contractors offering bids
- designers should be held accountable for the safety of their designs, much the same way as equipment and machinery designers, and
- contractors should consider safety at the tendering stage and at all stages of the construction project.

Finally, these Regulations are special because of the breadth of their application. The CDM Regulations apply to virtually all construction work. 'Construction work' is broadly defined to include most building and civil engineering work, with the exception of mineral extraction and exploration. For this reason, it is probably safest to assume that the Regulations apply to a project, whether it be new construction, repairs, maintenance, refurbishment or demolition.

What follows is a summary of the main duties created by the CDM Regulations. This summary is not intended to represent a complete statement of the Regulations. If detailed knowledge of the Regulations is required, you should refer in the first instance to the Regulations themselves, then to the Approved Code of Practice (ACoP) prepared for the Regulations by the Health and Safety Commission or to the Guidance Notes prepared by the Health and Safety Executive in conjunction with the Construction Industry Advisory Council (CONIAC). (See Chapter 1 for a discussion of the effect given to ACoPs and Guidance Notes.)

We will begin by considering the Health and Safety File and the Health and Safety Plan, the two fundamental components of the CDM Regulations' system. Thereafter, the duties of the various participants are outlined.

3.2.2. The Health and Safety Plan

The purpose of the Health and Safety Plan (the Plan) is to make plain to contractors bidding for a job, and to those working on the site, the health and safety issues specific to that project. For this reason the Plan is a document of two distinct stages: the pre-tender Plan and the final Plan.

The pre-tender Plan is to be organised by the Planning Supervisor (see Section 3.2.6. for details on the Planning Supervisor) appointed to the project. It should be prepared as soon as possible after the project is conceived and should form part of the tender documentation submitted to contractors tendering for the job.

The pre-tender Plan should set out all of the significant safety risks associated with the proposed project, thereby permitting the contractors to develop appropriate health and safety procedures and systems for the project.

To that end, it has been suggested that the pre-tender Plan contain, as a minimum

- any general information about the project supplied by the Client, including the projected completion date, site information, current uses and conditions, etc.

25

- details on the foreseeable health and safety risks identified in the designs
- construction methods recommended by the designers
- any other information that the Planning Supervisor knows or could reasonably ascertain after making enquiries that would assist contractors in proving their competence or the sufficiency of their health and safety resources, and
- any information that the Planning Supervisor believes the contractors would need to know to ensure the safety of their workers.

As mentioned, the pre-tender Plan must be included in the documentation provided to tendering contractors.

Once the Principal Contractor (see Section 3.2.7. below) has been appointed, it becomes his responsibility to develop the Plan to its final form. Most importantly, the Client must ensure that the final Plan has been satisfactorily developed before permitting construction work to begin.

When fully developed, the Plan should contain, as a minimum, information on

- the health and safety management rules and procedures developed for the site
- the welfare arrangements for the site
- the safety management structure developed for the project
- issues raised in the contractors' risk assessments prepared in accordance with the Management of Health and Safety at Work Regulations 1992, as amended (see Chapter 1 for a discussion of those Regulations)
- the rules for monitoring compliance with the Plan, and
- the rules for auditing the effectiveness of the Plan.

The Plan is to be updated and amended whenever appropriate, and may be prepared on a phase-by-phase basis if that is how the construction work is to proceed.

The ACoP to the Regulations suggests that any amendments made by the Principal Contractor to those parts of the Plan that were originally developed by the Planning Supervisor must be confirmed with the Planning Supervisor before being implemented.

3.2.3 The Health and Safety File

Like the pre-tender Plan, the Planning Supervisor has responsibility for preparing a Health and Safety File for every *structure* that comprises that construction project. That File should contain information on the structure's design, construction and how it is to be used by the occupants, so that any person planning for construction in the future will understand the hazards that may be associated with that structure.

To that end, the ACoP suggests that the following be included in the completed Health and Safety File, at a minimum

- record drawings and plans used throughout the period of construction
- general details of the construction methods and materials
- details of the equipment and maintenance facilities
- copies of maintenance procedures for the structure
- maintenance manuals and schedules for plant and equipment provided by contractors and suppliers
- details on the location and nature of utilities and services (e.g. emergency and fire fighting systems), and
- any information from the Health and Safety Plan that would be relevant for future projects.

3.2.4. The Client

The CDM Regulations are revolutionary among construction and safety rules in that they impose some responsibility in respect of safety on the site on the person contracting the work — the Client.

The 'Client' is any person for whom construction work is being carried out, whether done by external labour or in-house. Please note, however, that a householder who contracts to have work done on his own home is not considered a Client for purposes of the Regulations.

The following is a summary of the Client's duties under the CDM Regulations.

Table 4. Clients' duties under the CDM Regulations

The Client must

- appoint a Planning Supervisor and Principal Contractor who are competent and who have allocated adequate resources for health and safety
- provide information to the Planning Supervisor
- prohibit the start of construction work until the Health and Safety Plan is fully developed, and
- make the Health and Safety File available to any person who requires it or who acquires an interest in the site.

Appointment of Planning Supervisor and Principal Contractor

The CDM Regulations provide that the Client must appoint a Planning Supervisor and a Principal Contractor to the project at all times. The Principal Contractor must be a contractor working on the project and the ACoP recommends that he be either the main or managing contractor.

On the other hand the Planning Supervisor need not serve any other role on the project, and can be either an individual or a company, as appropriate. Also, appointment of more than one Planning Supervisor may be appropriate if, for example, the project is phased.

The Planning Supervisor and the Principal Contractor must be appointed as soon as appropriate. Having said that, in no event should the Client appoint the Principal Contractor or the Planning Supervisor until he has satisfied himself both of their 'competence' and that they have 'allocated adequate resources' to enable them to perform their functions. While the meaning to be given to those terms is far from precise, the ACoP suggests that when the Client reviews 'competence', he should consider, amongst other things

- the person's knowledge and understanding of the project at hand
- the person's familiarity with construction techniques generally
- their ability to manage risks, and

28

- their knowledge of health and safety.

Similarly, when the Client considers the adequacy of the resources of a candidate, the Client should consider

- the budget allocated by that person
- the number of persons scheduled to do the job, and
- the time allowed to complete the project.

In no event should a Client make any appointment until he has satisfied himself that these criteria have been met.

Provision of information
The Client is expected to provide information about the state and condition of the construction site to the Planning Supervisor as soon as practical before work commences. That information should include anything that would assist the Planning Supervisor in complying with his duties under the Regulations.

The Client and the Plan
Also, the Client must ensure, so far as is reasonably practicable, that an appropriate Health and Safety Plan has been developed by the Principal Contractor before permitting construction work to begin. The Plan will be deemed appropriate if it is sufficiently 'developed', i.e. it includes arrangements for ensuring the health and safety of all persons at work and others affected by the work. The ACoP makes it clear that the Client must give the Principal Contractor sufficient time to develop the Plan before construction is to begin.

The Client and the File
The Client must take reasonable steps to ensure that the information contained in the Health and Safety File is available for inspection by any person (e.g. subcontractors) who may need it to comply with his statutory duties. Ultimately, the Client must deliver the File (or a copy of the File) to any person acquiring an interest in the premises, such as a purchaser or a lessee.

3.2.5. The Designer

The 'Designer' is any person who prepares a design for construction, or who employs persons to prepare such designs. This broad definition would include architects and engineers, as well as quantity surveyors and project managers involved with design, for example.

Before a Designer can begin a design for a Client, he must take steps to notify the Client of the latter's duties under the CDM Regulations. In this way, it is hoped that the Client will undertake to fulfil his duties as quickly and effectively as possible.

Table 5 gives a summary of the Designer's duties under the CDM Regulations.

Table 5. Designer's duties under the CDM Regulations

The Designer must

- ensure that the Client is made aware of duties under the Regulations
- prepare risk assessment of designs
- provide information to Contractors on risks remaining in his designs, and
- co-operate with other Designers and the Planning Supervisor.

Designers' risk assessments

The essential part of the Designer's duties under the Regulations is to ensure, so far as is reasonably practicable, that at all stages in the life of a structure his design

- eliminates foreseeable risks to the health and safety of all persons performing the construction work, cleaning the structure, or who may be affected by the work of such persons
- combats at source all unavoidable risks to the health and safety of those persons, and
- gives priority to measures which protect all workers over measures that protect only one person.

The ACoP makes it clear that the use of the phrase 'so far as is reasonably practicable' permits the Designer to balance the risk of injury associated with a feature of the design against the costs of excluding that feature. Costs in this respect could include

- the 'buildability' of a design without that feature
- its fitness for purpose
- the aesthetics of the building, and
- the environmental impact of removing that feature.

In this way the CDM Regulations are requiring Designers to perform a risk assessment of their designs. When doing these assessments, however, the Designer may assume that

- his duty is only to eliminate or reduce reasonably foreseeable risks, and
- the persons building, maintaining and repairing the structure, as well as those persons affected, are competent.

(See Chapter 1 for a discussion of risk assessments.)

Design information

The Designer must ensure, so far as is reasonably practicable, that his designs include adequate information about the risks or injury associated with the project design and/or the materials to be used. As such, if aspects of the design need to be brought to the attention of the contractor, the Designer must provide that information. Having said that, the ACoP makes it clear that Designers are not expected to dictate construction methods to contractors.

Co-operation with others

The Designer must also co-operate with the Planning Supervisor and other designers to enable them to comply with their health and safety obligations. This co-operation could include exchanging information about aspects of his design when necessary to avoid or reduce health and safety risks.

The Designer must also inform the Planning Supervisor about aspects of the design that could present a risk of injury when interacting with designs prepared by others.

31

3.2.6. The Planning Supervisor

The 'Planning Supervisor' is the person responsible for co-ordinating and supervising the design phases of the project to ensure that the designs prepared adequately limit the health and safety risks, so far as is reasonably practicable.

Table 6 gives a summary of the Planning Supervisor's duties under the CDM Regulations.

Table 6. Planning Supervisor's duties under the CDM Regulations

The Planning Supervisor must

- advise the Client and others on appointments
- send notice of the project to the HSE
- prepare pre-tender Health and Safety Plan
- prepare Health and Safety File
- supervise the designers to ensure that they are fulfilling their statutory duties, and
- co-operate with the Principal Contractor.

Duty to supervise

The Planning Supervisor is expected to ensure that Designers fulfil their statutory duties to conduct risk assessments of their designs. Any review that the Planning Supervisor may undertake need only be to a standard that is 'reasonable for a person in his position to take'. As such, the Planning Supervisor is not expected to review every design or to redraw the designs, but rather to conduct a review of the procedures put into place by the Designers to conduct such assessments.

To the extent that the Planning Supervisor feels that the procedures put into place by the Designer are inadequate or could be improved, he may wish to review the designs himself, and/or make recommendations for changes to the design or the design review process.

Advise the Client and Contractors

The Planning Supervisor must provide advice, whenever requested to do so by the Client or any of the contractors, on the appointment of a

Designer, contractor or other consultant. The Planning Supervisor is required to give advice on that individual's health and safety competence and on the adequacy of his resources in terms of health and safety.

The Planning Supervisor's Plan

As previously mentioned, the Planning Supervisor has responsibility for ensuring the preparation of a pre-tender Health and Safety Plan. While the Regulations do not require the Planning Supervisor to prepare that Plan himself, he must ensure that such a Plan is prepared. As such, the Planning Supervisor may instruct another person to prepare the pre-tender plan provided that person has the requisite knowledge and information to perform the task effectively.

The Planning Supervisor's File

Finally, the Planning Supervisor must ensure that the Health and Safety File has been prepared during the construction project and that it is delivered upon completion to the Client. The difficulty lies in determining what is meant by the term 'completion'. As no definition of that term is provided in the Regulations, it is generally advisable to define that term in the contract with the Planning Supervisor.

3.2.7. The Principal Contractor

The 'Principal Contractor' will be a contractor working on the project who has been appointed by the Client to fulfil the duties listed in Table 7.

The following is a summary of the Principal Contractor's duties under the CDM Regulations.

Co-ordination duties

The Principal Contractor must take all reasonable steps to ensure co-operation between contractors, so that they all can comply with their statutory duties as regards health and safety.

Table 7. Principal Contractor's duties under the CDM Regulations

The Principal Contractor must

- co-ordinate the safety activities of the contractors on site
- restrict access to the site to authorised persons only
- develop the Health and Safety Plan
- reasonably direct the actions of contractors in respect of safety
- ensure that safety information and training are provided to all operatives working on the site about the risks and conditions
- obtain and consider the advice and opinions of persons on site
- provide information for the Health and Safety File, as appropriate, and
- obtain the advice and views of persons at work in respect of safety.

To the extent that contractors interact on the site, that interaction must be co-ordinated by the Principal Contractor in the manner outlined in the Health and Safety Plan. The Principal Contractor must also lead in the preparation and communication of co-ordinated emergency procedures.

Supervision duties

The Principal Contractor must supervise and assess the performance of contractors, so far as is reasonably practicable, to ensure that they are complying with the health and safety rules developed for the site.

Limiting site access

It is the Principal Contractor's responsibility to ensure that only 'authorised persons' are allowed on to the construction site. 'Authorised persons' are those persons who have been invited on to, or who have a statutory or contractual right to enter all, or part, of the construction area. The Principal Contractor must take all reasonable steps to exclude all others from the work area. How this is done is dependent upon the nature of the project, but the measures need only address access that is foreseeable.

As such, if a trespasser takes extraordinary measures to enter a site, the Principal Contractor will not necessarily be seen as failing in his duty to prevent that access. The ACoP provides that on large remote sites, warning signs may be sufficient, whilst sites in developed areas may require more permanent security measures.

Display notification

A copy of the notice sent to the HSE must be displayed at the site by the Principal Contractor in places and in a condition where it can be easily seen and read by all. Generally, this will be at the site entrance, at a permanent place on the perimeter or at the site office. On large sites the notice may need to be displayed at several locations.

Direction of contractors

The Principal Contractor has the authority to *reasonably* direct the actions of contractors whenever necessary to ensure compliance with the health and safety rules developed for the site. Correspondingly, contractors must comply with the reasonable directions given by the Principal Contractor as regards health and safety.

Development of the Health and Safety Plan

As discussed in Section 3.2.2., it is the Principal Contractor's responsibility to 'develop' the Health and Safety Plan once he has been appointed. The developed Plan must be completed before the Client can permit construction to begin. It is the further responsibility of the Principal Contractor, after consultation with the contractors, to monitor the effectiveness of the Health and Safety Plan and to make modifications as necessary.

Provision of Information and Training

The Principal Contractor must ensure, so far as is reasonably practicable, that every contractor is provided with information on the risks to the health and safety of that contractor's employees. Furthermore, the Principal Contractor must supply the contractor with all information that he needs to comply with his duty as an employer under the Management of Health and Safety at Work Regulations 1992, as amended, to provide adequate information and training to his employees. (See Chapter 1 for a discussion of those Regulations.)

Advice and views of persons at work
Finally, the Principal Contractor must ensure that all employees are consulted on the health and safety arrangements for the site. To do this, he must consult with the safety representatives and safety committees, or make other arrangements for communicating with employees in the event that safety representatives have not been appointed.

3.2.8. The Contractor

A 'Contractor' is any person who carries out or manages construction work or who organises others to carry out such work on his behalf (e.g. subcontractors or works package contractors). For this reason, it may be the case that a project manager or works supervisor is treated as a Contractor for purposes of the CDM Regulations.

Contractor's duties
For the most part, the Contractor's duties under the CDM Regulations simply augment the duties he has under other health and safety legislation. In particular, the Contractor must

- co-operate with the Principal Contractor to the extent necessary for the latter to comply with his duties
- provide the Principal Contractor with any information, so far as is reasonably practicable, which might affect the health and safety of any person on the site or which might affect the validity of the health and safety arrangements for the site
- comply with reasonable directions given by the Principal Contractor for the purpose of improving health and safety on the site
- inform the Principal Contractor of any injuries, accidents or dangerous occurrences, so that he can monitor the effectiveness of the health and safety arrangements
- comply with all health and safety rules for the site, and

- provide the Principal Contractor with such information as he reasonably believes should be included in the Health and Safety Plan.

3.3. EXCLUSION OF CIVIL LIABILITY

The CDM Regulations provide that a breach of a duty contained in the Regulations does not confer a right on a plaintiff to bring a civil action for damages associated with a breach of statutory duty, with two exceptions. Those exceptions are

- the Client's failure to prevent construction work to begin until a Health and Safety Plan has been prepared and developed, and
- a Principal Contractor's failure to take reasonable steps to ensure that only authorised persons are allowed on to the construction site.

Therefore, if a person is injured as a result of one of these breaches, he may bring a civil action for damages, alleging breach of statutory duty. (See Chapter 1 for a discussion of an action for breach of statutory duty.)

4

Equipment safety

4.1. INTRODUCTION

Undeniably, the use of equipment and machinery presents a significant health and safety risk on a construction site. For that reason it is not surprising that there has been a large amount of legislation on this issue.

This Chapter will consider equipment safety in general terms and then consider legislation associated with specific types of work equipment, including cranes, hoists and access equipment generally. Finally, we outline the requirements of the new Lifts Regulations 1997.

4.2. EQUIPMENT SAFETY — GENERALLY

4.2.1. The Provision and Use of Work Equipment Regulations 1992

All equipment and machinery provided for use at a work site must comply with the general requirements of the Health and Safety at Work Act 1974, and in particular s. (2)(a), which obliges employers to provide their employees with safe plant and equipment. In addition, employers are obliged to comply with the requirements of the Provision and Use of Work Equipment Regulations 1992 (the Work Equipment Regulations), which specifically address the health and safety risks associated with work equipment. The following is a summary of those Regulations, with particular emphasis given to issues affecting the construction industry.

Application of Regulations
The Work Equipment Regulations set standards for the equipment which is provided by employers for use in the workplace. In addition, the Regulations apply to work equipment used by self-employed persons, and to equipment that is provided by persons in control of premises. Given the breadth of these application rules, it may be a good idea for employers to assume that all work equipment on the site is covered by these Regulations — whether supplied by the employer or brought on to site by others.

Suitability of work equipment
The Regulations require employers to ensure that equipment provided is suitable for the purpose for which it is being used and the task for which it has been provided. In making that decision, the employer must consider the working conditions in the area in which the equipment is to be used.

Maintenance
All work equipment is to be maintained in good working order at all times. In addition, records must be kept of such maintenance.

Information, instruction and training
Employers must provide their employees with information and training on the health and safety risks associated with the work equipment that they will be using and on the proper methods for using such equipment. Similarly, employers must ensure that equipment is only

used by those persons who are trained to use it and repaired only by those persons trained to repair it.

Equipment standards

All equipment provided for use in the workplace must comply with relevant European safety standards. In the UK, this will mean complying with the Supply of Machinery (Safety) Regulations 1992 at the very least, as well as any machinery-specific standards. (For a detailed discussion of the Supply of Machinery (Safety) Regulations, see Section 4.2.3.)

Equipment guards

Employers must ensure that access to all dangerous parts of equipment is either prevented or that movement of any such part is halted before a person could come into contact with it. This obligation should be satisfied by the use of (in order of priority)

- fixed guards
- other guards or other protection devices
- jigs
- push-sticks, or
- other similar protection.

Only when none of these methods is practicable can the employer rely simply on information, instruction, training and supervision.

Controls

Employers must ensure that all controls on equipment or machinery are sufficient to

- stop the equipment completely in the event of an emergency, and
- switch off all energy supplies to equipment in the event of an emergency.

Employers must also ensure that those controls

- are clearly visible and marked

41

- are located in a safe and convenient location
- do not expose the operator to risks when used, and
- are part of a safe control system, which does not by its operation permit faults or damage that will increase the risk of injury.

Other issues

All equipment must be equipped with suitable means for isolating energy supplies, which means shall be clearly identifiable and accessible.

Stability must be assured for all equipment and all equipment must be placed in an area with suitable lighting.

Every employer shall ensure that equipment maintenance can take place while the equipment is shut down, or alternatively done without risk to safety.

All equipment must be appropriately marked, with appropriate safety warnings attached.

4.2.2. Proposed Work Equipment Regulations

In July 1997, the HSE proposed several new regulations in respect of work equipment. When adopted, these new laws will amend the existing health and safety requirements in respect of

- non-lifting equipment
- lifting equipment
- woodworking equipment, and
- power presses.

These proposals will also implement in the UK a new European Council Directive on work equipment, which was adopted in 1995. At the time of writing, these new regulations are set to come into force by 1998.

4.2.3. Supply of Machinery (Safety) Regulations

In addition to the requirements of the Work Equipment Regulations, employers and others who put machinery into service must comply with the Supply of Machinery (Safety) Regulations 1992, as amended by the Supply of Machinery (Safety) (Amendment) Regulations 1994 (collectively, the Supply Regulations). The following is a brief summary of those Regulations.

Suppliers' duties
The Supply Regulations impose duties on suppliers of machinery and machinery components. A machinery supplier is any person who puts machinery into service in the course of a business for the first time. Given the breadth of that definition, these Regulations may affect employers in the construction industry any time they put a new piece of equipment or machinery into service on a site.

Before putting machinery into use, suppliers must ensure first that it meets the essential health and safety requirements for machinery of that type. Essential safety requirements cover matters such as

- the materials used to construct the machinery
- the inclusion of integral lighting on the machinery as appropriate
- handling systems for the machinery
- measures taken to prevent slipping and tripping near machinery
- the adequacy of controls and control systems
- guards and other hazard protections
- maintenance systems, and
- the warnings, markings and instructions for use that are provided.

Machinery suppliers are advised to refer to the manufacturer's standards when determining if the machinery satisfies the essential safety standards. Furthermore, it may be necessary to consult any relevant safety standards for equipment of that type, as appropriate.

The Supply Regulations also oblige suppliers to ensure that the appropriate 'conformity procedures' have been carried out on the machinery. These procedures might require independent testing by a responsible person, who is either

- the manufacturer of the machinery
- the manufacturer's authorised representative in the European Community, or

- any person who supplies the machinery into the Community.

Alternatively, the machinery supplier may ensure that the machinery has been issued with either a declaration of conformity or a declaration of incorporation, as appropriate. However, in all cases a 'CE' mark must be properly affixed to the machinery by the appropriate person. The CE mark is intended to signify that a piece of machinery is indeed safe to use.

For further information about these requirements, the reader is advised to refer to the Department of Transportation's publication entitled *Guidance Notes on the Supply of Machinery (Safety) Regulations*.

4.3. LIFTING EQUIPMENT

In addition to the duties imposed by the HSW Act, the Work Equipment Regulations and the Supply Regulations, employers who offer lifting equipment for use by their employees must also comply with legislation specific to such equipment, which includes the Construction (Lifting Operations) Regulations 1961 (the Lifting Operations Regulations) and the Factories Act 1961. Together these statutes set out in detail the standards that lifting equipment must meet. (See Table 8 for a list of lifting equipment covered by these statutes.)

We will begin with a summary of the Lifting Operations Regulations and then consider the specific requirements that have been developed for various types of lifting equipment. Insofar as the term 'lifting appliances' covers a wide range of equipment and a wide range of uses, we have elected to break such equipment into two main types — namely cranes and hoists.

Please note, however, that these standards may be subject to change soon, in view of the proposals put forward by the HSE in July 1997 to amend the existing regulations covering lifting equipment. These proposals are contained in the HSE Consultative Document (CD116) entitled *Proposals for the Implementation of the Lifting Aspects of the Amending Directive to the Use of Work Equipment Directive*.

4.3.1. Construction (Lifting Operations) Regulations 1961

For the most part, the Lifting Operations Regulations mirror the requirements of the Factories Act with respect to lifting equipment. Having said that, the Lifting Operations Regulations go into much more prescriptive detail.

For example, the Factories Act prescribes that all lifting equipment must be of good construction, based on sound material, properly maintained, of adequate strength and free from defects, regularly inspected and marked with the safe working load (SWL) for that equipment. The Lifting Operations Regulations, on the other hand, prescribe that lifting equipment must be

- of good mechanical construction, sound material, adequate strength and free from patent defect
- properly maintained
- as far as the construction permits, inspected at least once in every week by a competent person, who shall record and keep his findings.

Support for all lifting appliances must be adequate and secure and of good construction and adequate strength.

The Regulations also set out standards for the size and construction for

- drivers' platforms
- drivers' cabins
- drums and pulleys
- brakes, controls and safety devices
- means of access to a lifting appliance, and
- the support for pulleys and gin wheel.

All lifting equipment must be regularly tested, inspected and/or examined by a 'competent person'. A competent person is any person who has been trained in the use and operation of that lifting appliance. Please note that the testing requirements have been supplemented by the Construction (Health, Safety and Welfare) Regulations 1996, which are described in detail in Chapter 5.

4.3.2. Cranes

Naturally specific types of lifting equipment have specific safety standards, and cranes are no exception. The standards developed for cranes emphasise the fact that failure to use a crane in accordance with its limitations and its SWL can, and often does, lead to accidents and injuries. For that reason both the Factories Act and the Lifting Operations Regulations lay out a number of standards for the use and construction of cranes.

Factories Act Requirements for Cranes

The Factories Act sets out a number of requirements specifically in respect of cranes. These include that a crane must be

- of good construction
- free from patent defects
- of adequate strength, and
- properly maintained.

Similarly crane rails and tracks must be

- of proper size
- of adequate strength
- on an even surface
- properly laid
- adequately supported or suspended, and
- properly maintained.

Finally, the SWL must be marked on every crane, crab, winch, pulley, gin wheel, derrick pole or mast and every aerial cableway and must never be exceeded — except in testing.

Lifting Operations Regulations requirements for cranes

For the most part, the standards set out in the Lifting Operations Regulations for cranes mirror the standards set out in the Factories Act. However, the Regulations also add a few more standards, including that

- cranes must be securely anchored or adequate ballasting must be used whenever raising or lowering materials
- all cranes with a derricking jib must be fitted with an effective interlocking device, and
- cranes are not to be used for any purpose other than to vertically raise or lower, unless the movement does not cause undue stress or instability.

Together, the Lifting Operations Regulations and the Factories Act establish a comprehensive set of design and construction standards for cranes.

Testing requirements for cranes
Both the Factories Act and the Lifting Operations Regulations contain precise testing and examination requirements for cranes and related equipment. The main testing requirements are listed in Table 8.

Copies of all examination and tests reports must be retained in accordance with the provisions of the Lifting Plant and Equipment (Record of Test and Examination, etc.) Regulations 1992. Employers have the right to use their own forms or can use Forms 2530 or 2531 developed by the HSE, as appropriate. Copies of those forms are included as Figures 1 and 2, respectively. The HSE has also published guidance for employers on these Regulations, entitled *A Guide to the Lifting Plant and Equipment (Record of Test and Examination, etc.) Regulations* (L20).

Additional guidance on cranes
The HSE has also prepared a number of publications on the use of cranes on construction sites, including a construction information sheet entitled *Safe Use of Mobile Cranes on Construction Sites* which suggests that the main causes of accidents and dangerous occurrences involving cranes are inadequate training, planning and maintenance.

Table 8. Lifting equipment — testing and examination requirements

Legislation	Type of equipment covered	Testing/examination frequency
The Factories Act 1961	All lifting machines.	All lifting equipment must be tested and examined by a competent person before being put into use and retested once every 14 months thereafter.
		Hoists must be examined by a competent person once every six months, except for continuous-use hoists and lifts, which must be examined every twelve months.
Construction (Lifting Operations) Regulations 1961	All lifting appliances, including crabs, winches, pulley blocks or gin wheels used to raise or lower, hoists, cranes, shear legs, excavators, piling frames, draglines, aerial cableways, aerial ropeways and overhead runways. All lifting gear, including chain slings, rope slings, or other similar gear and a ring, link hook, plate clamp, shackle, swivel or eye-bolt.	Cranes, crabs and winches must be tested and examined by a competent person once every 48 months, and after every major repair or alteration. Pulley blocks, gin wheels and sheer legs must be tested and examined by a competent person before being put into use, and after every major repair or alteration. All lifting appliances (except hoists) must be tested and examined by a competent person every 14 months, and after every major repair or alteration. Appliances for anchoring and/or ballasting a crane must be examined by a competent person before every use and tested once the crane has been erected. Automatic safe load indicators must be tested by a competent person every week.

HSE
Health & Safety
Executive

Record No. _____

RECORD OF THOROUGH EXAMINATION OF
LIFTING PLANT AND EQUIPMENT

Description of equipment

Identification mark of the equipment

Name and address of owner of equipment, and its location

Date of the most recent test and examination or test and thorough examination

and date and number or other identification of the record issued on that occasion

Date of last thorough examination and number or other identification of the record issued

Safe working load or loads and (where relevant) corresponding radii

Details of any defects found (if none state NONE)

Date(s) by which defects described above must be remedied

Any other observations

What parts if any, were inaccessible?
(TO BE COMPLETED ONLY AFTER A THOROUGH EXAMINATION OF A HOIST OR LIFT)

Latest date by which the next thorough examination must be carried out

Declaration
I hereby declare that the equipment described in this record was thoroughly examined in accordance with the appropriate
provisions and found free from any defect likely to affect safety other than those listed above on (date)...............................
and that the above particulars are correct.

Signature or other identification

Name and address of person authenticating the record

Date the record is made

Name and address of person responsible for the thorough examination

F2530 (8 / 95)

Figure 1. HSE Form 2530

49

HSE
Health & Safety
Executive

Record No. _____

RECORD OF TEST, TEST AND EXAMINATION OR TEST AND THOROUGH EXAMINATION OF LIFTING PLANT AND EQUIPMENT

Description of the equipment
(including date of manufacture)

Name and address of owner of equipment, and its location

Identification mark of the equipment

Safe working load or loads and (where relevant) corresponding radii, jib lengths and ... ...ghts

Details of the test, test and examination or test and thorou... ...an... ...io...arried out

Date or dates of completion

Declaration

I hereby declare that the equipment described in this record was tested, tested and examined or tested and thoroughly examined in accordance with the appropriate provisions and is found free from any defect likely to affect safety on (date)... and the above particulars are correct.

Signature

Name and address of person making above declaration
(Typed or printed)

Date the record is made

F2531 (8 / 95)

Figure 2. HSE Form 2531

50

That sheet also provides guidance on

- selecting the correct crane to use
- maintaining the proper paperwork and records
- proper placement and siting of the crane
- considerations as to the type and weight of the load
- the maintenance and repair of lifting gear and equipment, including slings, wire ropes and webbing
- managing the lift, including planning and monitoring the lift
- proper use and standardisation of signals
- maintenance and inspection of crane and related equipment, and
- installation of an automatic safe load indicator to all cranes capable of lifting more than one tonne.

Finally, the collapsing or overturning of a crane is a reportable incident under the Reporting of Injuries, Diseases and Dangerous Occurrences Regulations 1995. Please see Chapter 9 for details on those Regulations.

4.3.3. Hoists

Like cranes, hoists are regulated by both the Factories Act and the Lifting Operations Regulations, which are summarised below.

Factories Act requirements for hoists
The Factories Act specifies that hoists must be of sound construction, of adequate strength for the task and properly maintained. In addition, every hoist must

- be fitted with a gate
- prevent unnecessary falls, and
- prevent unnecessary contact with moving parts.

All hoist gates must be interlocking so that they can be opened only when the hoist is at an opening — not during movement.
Further, to the extent that a hoist is used to move people, it must

- be fitted with an automatic device to prevent it running over
- be fitted with an interlocking device to prevent the gate from being opened except at teagle openings, and to prevent the hoist from moving when the gate is opened, and

51

- for suspension hoists, be fitted with at least two ropes, chains or cables, each capable of carrying the maximum load.

Lifting Operations Regulations requirements for hoists
The Lifting Operations Regulations sets out additional safety requirements for hoists, including that

- every moving part of the hoist or any hoistway is protected by a substantial enclosure
- where practicable, the hoist should be operable from one position
- signalmen should be used when the hoist operator has an obstructed view, and
- the SWL of the hoist be stated on the cage.

There are numerous exemptions to the application of these rules, which are listed in the Hoists Exemption Order 1962, as amended by the Hoists Exemption (Amendment) Order 1967.

Testing requirements for hoists
The examination and testing requirements for hoists are set out in Table 8. As for cranes, copies of those examination and test reports must be maintained in accordance with the Lifting Plant and Equipment (Record of Test and Examination, etc.) Regulations 1992, which are discussed in Section 4.3.2.

Miscellaneous guidance on hoists
The HSE has produced extensive guidance for employers on the proper use and construction of hoist, which is listed in the Bibliography.

4.4. ACCESS EQUIPMENT

The safety of access equipment is a fundamental part of safety on a site, as access equipment is a fundamental part of every construction project, and as such has been the subject of a great amount of statutory treatment — and in particular, a great deal of guidance from the HSE.

4.4.1. Statutory requirements

Unlike lifting equipment, access equipment is not specifically considered in the Factories Act. However, it is mentioned in the Lifting Equipment Regulations. In particular, those Regulations require that all access equipment (including scaffolding, framework, suspended access, cradles, etc.) be

- of good construction
- of adequate strength
- made of sound materials, and
- free from defect.

This obligation is echoed in Regulation 7 of the Construction (General Provisions) Regulations 1961, which provides that if an employer provides either scaffolding, ladders, or other means of access, he must ensure that such equipment is suitable and sufficient for the task.

These requirements have been supplemented by those provisions of the Construction (Health, Safety and Welfare) Regulations 1996 that deal with fall prevention, which provisions are described in detail in Chapter 5.

Lastly, those who design and/or supply scaffolding for access on a construction site are subject to the requirements of the Construction (Design and Management) Regulations 1994, which are described in detail in Chapter 3.

4.4.2. Access equipment guidance

To supplement the statutory requirements, the HSE has produced a number of guidance notes for employers on a variety of access equipment, such as access scaffolds, tower cranes, suspended access equipment, etc. This guidance includes titles such as *Scaffold Towers*,

Suspended Access Equipment, and *General Access Scaffolds*. A complete list of publications is included in the Bibliography.

4.5. LIFTS REGULATIONS

On 1 July 1997, the Lifts Regulations 1997 came into effect. These Regulations set out minimum safety standards for lifts which service buildings and construction sites. 'Lifts' are defined to include all appliances which incorporate, for example, a car to move persons and/or goods along a fixed line; however, there are a number of exceptions to that rule.

5

Workplace safety

5.1. INTRODUCTION

Conditions in the workplace are fundamental to ensuring the health and safety of the people working there. While that maxim is true for any workplace, it is particularly true on construction sites, in view of the inherently dangerous nature of the work and the fact that several teams of workers are all working together under one 'roof'.

For this reason construction worksite safety legislation underwent a substantial overhaul in 1996 in the UK with the introduction of the Construction (Health, Safety and Welfare) Regulations 1996. These Regulations revoke many site safety laws that had been in place since the 1960s. In view of the fundamental nature of those Regulations, we will consider them in some detail in this Chapter.

In addition, certain fixed premises used in connection with a construction or engineering project (e.g. off-site accommodation or offices) are subject to another set of workplace requirements, namely the Workplace (Health, Safety and Welfare) Regulations 1992. In view of their possible application to a construction project, those Regulations will also be outlined in this Chapter.

Finally, this Chapter will consider a few individual regulations that address two specific workplace issues, namely lighting and noise.

5.2. THE CONSTRUCTION (HEALTH, SAFETY AND WELFARE) REGULATIONS 1996

When they came into force, the Construction (Health, Safety and Welfare) Regulations 1996 (the Construction HSW Regulations) revoked several long-standing construction safety laws, including

- Regulations 8 to 19, 21, 23 to 41, 45 to 51, 53 and 56 of the Construction (General Provisions) Regulations 1961 (SI 1961/1580)
- all of the Construction (Working Places) Regulations 1966, and
- all of the Construction (Health and Welfare) Regulations 1966.

The Construction (HSW) Regulations also implement Annex IV of the Temporary or Mobile Construction Sites Directive, which is the Directive that led to the creation of the Construction (Design and Management) Regulations 1994, which are described in detail in Chapter 3.

For these reasons, these Regulations represent a fundamental part of the construction safety regime.

5.2.1. Application of the Regulations

Like the Construction (Design and Management) Regulations 1994, the Construction HSW Regulations apply to all sites used for construction work. 'Construction work' is defined quite broadly in both sets of regulations to include most building, civil engineering or engineering construction work. Table 9 provides a complete definition of 'construction work'.

The duties set out in the Construction HSW Regulations are imposed on employers, self-employed persons, persons in control of the way others work and every employee carrying out construction work.

Table 9. Definition of 'Construction Work' in the Construction HSW Regulations

Construction work includes

- the construction, alteration, conversion, fitting out, commissioning, renovation, repair, upkeep, redecoration or other maintenance (including cleaning which involves the use of water or an abrasive at high pressure or the use of substances classified as corrosive or toxic... decommissioning, demolition or dismantling of a structure
- the preparation for an intended structure, including site clearance, exploration, investigation (but not site survey) and excavation, and laying or installing the foundations of the structure
- the assembly of prefabricated elements to form a structure or the disassembly of prefabricated elements which, immediately before such disassembly, formed a structure
- the removal of a structure or part of a structure or of any product or waste resulting from demolition or dismantling of a structure or from disassembly of prefabricated elements which, immediately before such disassembly, formed a structure, and
- the installation, commissioning, maintenance, repair or removal of mechanical, electrical, gas, compressed air, hydraulic, telecommunications, computer or similar services which are normally fixed within or to a structure

Construction work does *not* include the exploration for or extraction of mineral resources or activities preparatory thereto carried out at a place where such exploration or extraction is carried out.

5.2.2. Safe place of work

The Regulations specify that, so far as is reasonably practicable, every construction site must

- provide safe and suitable access and egress
- be made safe for the people who work there, and
- provide sufficient working space for any person likely to work on that site.

This latter duty reinforces the general duty to provide a safe place of work, as set out in s. 2 of the HSW Act.

5.2.3. Fall prevention

Appropriate steps must be taken to prevent any person from falling on the construction site. In particular, if a person is likely to fall two metres or more, suitable and sufficient toe-boards, guard-rails, barriers, work platforms or some other similar means must be provided, so far as is reasonably practicable.

Any scaffolding or other protective equipment that is erected as a means for preventing falls must be erected under the supervision of a competent person. The term 'competent person' is used repeatedly throughout the Regulations and refers to a person with the necessary training and/or experience to ensure that a particular operation is performed correctly and safely.

The Regulations also prohibit the use of a ladder as a means of access/egress to a place of work unless it would be reasonable to use a ladder in view of the nature or duration of the work and the limited risks to the person using it.

In particular, appropriate steps should be taken to prevent any person from falling through fragile material. If there is a chance of such a fall, that person shall be provided with the correct equipment to arrest any falls and notices must be put up warning them of the dangers. Only when it is not reasonably practicable to provide means to prevent falls will it be appropriate for an employer to rely on the provision of personal protective equipment as the principal means of providing protection. (See Chapter 7 for a discussion of the personal protection equipment requirements.)

Any equipment provided to limit the risk of injury from falling must be properly maintained and must comply with the relevant requirements of the Regulations, which are set out in the Schedules thereto.

5.2.4. Falling objects

Suitable and sufficient efforts must be taken to prevent the fall of any object or material, so far as is reasonably practicable. Those efforts shall include the provision of

- toe-boards

- guard rails
- barriers, or
- a working platform.

When it is not reasonably practicable to provide any of these measures, then appropriate steps must be taken to prevent injury from falling material.

5.2.5. Stability of structures

Steps must be taken to prevent structures from being weakened by ongoing construction work or by overloading. A competent person must supervise the erection or dismantling of any structural supports.

5.2.6. Demolition or dismantling

Suitable and sufficient steps must be taken to prevent the risk of injury from the demolition or dismantling of any structure. Furthermore, all demolition or dismantling work must be supervised by a competent person.

5.2.7. Explosives

Suitable and sufficient steps must be taken to prevent the risk of injury from flying material associated with the use of explosives.

5.2.8. Excavations

All practicable steps (please note that this section is not limited by the word 'reasonably') must be taken to prevent danger presented by weak or unstable excavated areas. Further, all reasonably practicable steps must be taken, under the supervision of a competent person, to prevent any person from being trapped or buried by fallen or dislodged materials, which steps could include supporting the walls of the excavated area.

Also, suitable steps must be taken to prevent any person, vehicle, equipment or materials from falling into any excavated area. No equipment or vehicles are to be stored at or near the edge of any excavated area without proper support.

5.2.9. Vehicles and traffic routes

Both pedestrian and vehicular traffic on a construction site must be organised safely, which organisation should include

- traffic routes of appropriate size and location
- adequate space between pedestrian doors/gates and vehicular traffic routes
- use of warning systems notifying pedestrians of oncoming traffic, and
- at least one pedestrian exit point in every loading bay.

All traffic routes must be kept free of obstruction, so far as is reasonably practicable, and have all appropriate safety signs. Vehicles should be driven, loaded and unloaded in a safe fashion.

5.2.10. Emergency procedures

Suitable and sufficient steps shall be taken to limit the risks of injury associated with fire, explosion, flooding or asphyxiation, which steps should include the proper identification of and maintenance of all emergency routes and exits. Traffic routes shall be lit at all times, preferably by natural lighting. Emergency lighting shall be made available on emergency routes, at exits and wherever necessary to limit the risk of injury associated with inadequate lighting.

In addition, employers must develop emergency procedures for any foreseeable emergency. Those may include evacuation procedures in the event of, for example, natural disasters or bomb threats. Those procedures must be practised and tested at appropriate intervals.

Fire-fighting equipment, fire detectors and alarms must be provided for all construction sites as necessary, and should be maintained and tested at regular intervals. Every person working on a construction site must be trained in the use of that equipment and with fire prevention techniques generally, so far as is reasonably practicable. (See Chapter 6 for more information on fire safety requirements.)

5.2.11. Welfare

Any persons in control of a construction site, as well as every employer and/or self-employed person on that site, must ensure that the following welfare facilities are provided for the use of the employees on the site

- adequate sanitary conveniences
- washing facilities (including showers if required)
- an adequate supply of drinking water, located at accessible and suitable places
- clothing accommodation areas
- changing areas, as appropriate, and
- rest facilities, in particular non-smoking areas that are separated from smoking areas.

5.2.12. Training

Every person must be given suitable and sufficient safety information, instruction and training for the work that they are asked to perform on site. This duty is a restatement of the general duty to provided information and training found in s. 2 of the HSW Act, which is described in detail in Chapter 2.

5.2.13. Inspection

All working platforms, suspension equipment, excavation sites, cofferdams and caissons must be inspected by a competent person in accordance with the procedures laid out in the Schedules to the Construction HSW Regulations.

To the extent that an inspection identifies any unsatisfactory conditions, those conditions must be remedied before the platform, excavation, etc., is put into use. A report must be made of every inspection of working platforms or other means of support, which report must retained for at least three months.

5.2.14. Miscellaneous requirements

- All cofferdams and caissons must be of suitable design and construction for their purpose.
- Whenever there is a chance of drowning associated with an excavated area on site, suitable and sufficient steps must be taken to prevent drowning.
- Whenever a door or gate is supplied for the purpose of limiting the risk of injury, it shall be fitted with appropriate safety devices.
- A supply of fresh or purified air should be provided to all persons on the site, so far as is reasonably practicable, and devices shall be fitted to any system used to provide that air warning of defects.
- Every part of the construction site must be kept clean and in good order, with the site perimeters marked with appropriate safety signs, so far as is reasonably practicable.

5.3. THE WORKPLACE (HEALTH, SAFETY AND WELFARE) REGULATIONS 1992

The Workplace (Health, Safety and Welfare) Regulations 1992 (the Workplace Regulations) apply to all workplaces *except* workplaces where building operations or engineering works are being undertaken. As such, the Workplace Regulations will apply to premises used by construction personnel to the extent that they are *off* the construction site. Examples of such fixed premises might include permanent office blocks, accommodation, off-site restaurant facilities, conference rooms, etc. On the other hand, on-site offices must comply with the construction HSW Regulations.

5.3.1. Maintenance and cleanliness of workplace

Every workplace must be maintained and cleaned in an efficient state and in good working order. Where appropriate, a suitable equipment maintenance system shall be arranged.

5.3.2. Ventilation, temperature and lighting

Adequate ventilation shall be provided for all workplaces, with a sufficient quantity of fresh or purified air being provided at all times. The systems used to ensure such ventilation must be fitted with warning devices in the event of failure.

Similarly, the temperature shall be maintained at a reasonable level for all workplaces from systems that do not permit the introduction of harmful fumes, gases or vapours.

Employers must ensure that suitable and sufficient lighting (and preferably natural lighting) is provided in every workplace, which may require the provision of emergency lighting, as appropriate.

5.3.3. Space and workstation requirements

Every room where persons work must have adequate floor area, height and space for ease of movement and comfort. The Guidance to the Workplace Regulations suggests that the space provided be at least eleven cubic metres per person in every room.

Similarly, every workstation provided for use must be suitable for the persons who are to use it, and must, at a minimum

- have a suitable seat and a footrest, as appropriate
- be protected from adverse weather, so far as is reasonably practicable
- easy to leave, in the event of any emergency, and
- designed to avoid slips or falls.

5.3.4. Fall prevention

Suitable measures must be taken, so far as is reasonably practicable, to prevent

- any person falling from a distance likely to cause injury
- persons being struck by falling materials.
- persons falling into any tank, pit or other structure containing a dangerous substance.

5.3.5. Floors and traffic routes

Floor and traffic areas must be constructed of suitable materials. In addition, pedestrian and vehicular traffic must be segregated as much as possible so that both can travel safely through the workplace. Safety signs shall be provided, as necessary, to indicate the traffic routes established.

Any moving walkways provided must also function safely and be equipped with appropriate safety devices.

5.3.6. Windows, doors, gates, walls and skylights

All windows, doors, gates, etc. shall be constructed of suitable materials. For example, transparent doors or windows may need to be constructed from non-breakable materials, to the extent that safety requires it. The Construction HSW Regulations specify a number of construction requirements for doors and gates, including tracking devices and other fail safe features.

5.3.7. Other welfare facilities

Suitable arrangements must be provided on the site for

- sanitary conveniences
- washing facilities
- drinking water supplies
- accommodation areas for clothing
- changing facilities, and
- facilities for eating and/or resting.

As this list clearly demonstrates, the duties imposed on employers, employees and the self-employed by the Workplace Regulations mirror the requirements of the Construction HSW Regulations to a large degree.

5.4. OTHER WORKPLACE STANDARDS

5.4.1. Lighting

There are a number of statutory instruments which impose duties on employers in respect of lighting. These statutes are designed to limit the risk of injury to workers' eyesight and/or to reduce the chances of poor lighting creating potentially dangerous conditions. The following is a brief summary of some of that legislation.

The Electricity at Work Regulations 1989 and the Provision and Use of Work Equipment Regulations 1992
Both sets of Regulations impose duties on employers and the self-employed to provide adequate lighting for work areas and work equipment, when necessary to prevent or reduce the risk of injury in view of the type of work being performed.

These duties are largely a restatement of those contained in the Workplace Regulations, as discussed in Section 5.3. above.

The Health and Safety (Display Screen Equipment) Regulations 1992
Whenever display screen equipment (e.g. visual display units and personal computers) is used on a construction project, employers must ensure that satisfactory lighting is provided and that appropriate contrast levels exist between the screen and the workstation. (See Chapter 11 for a detailed discussion of these Regulations.)

The Manual Handling Operations Regulations 1992
These Regulations oblige employers to assess the adequacy of lighting conditions in the workplace before requiring a manual operation to be undertaken. (See Chapter 11 for a detailed discussion of these Regulations.)

HSE Guidance: 'Lighting at Work'
This Guidance Note provides assistance and advice to employers and others on the effects of lighting on the health and safety of persons at work. Specifically it offers information on the effects of illuminance,

glare and colour, and gives recommendations on lighting in a variety of workplaces, including construction sites.

5.4.2. Workplace noise

As with lighting, the level and/or type of noise on a construction site can have a fundamental effect on safety — both in terms of the damage it can cause to workers' hearing and because it may make it difficult for persons to hear warnings or alarms. The following is a brief summary of some of the relevant legislation relating to noise.

The Noise at Work Regulations 1989

These Regulations oblige employers and self-employed persons to

- assess the likelihood that persons could be exposed to excessive noise levels
- record the results of that assessment
- reduce the risk of hearing damage to the lowest levels reasonably practicable
- provide hearing protection when noise levels could exceed 85 decibels dB(A)
- establish hearing protection zones when noise levels could exceed 90 decibels dB(A)
- designate hearing protection zones through use of appropriate safety signs
- ensure that hearing protection equipment is used and maintained properly
- provide information and training on the use of hearing protection equipment and on the risk of injury associated with noise.

The Supply of Machinery (Safety) Regulations 1992

The designers and manufacturers of machinery that is used on construction sites must reduce the risk of injury associated with machinery noise to the lowest levels possible. In addition, machinery manufacturers must supply information to users as to the levels of noise emitted by that machinery, which information shall include instructions for reducing noise or vibration. (See Chapter 4 for a detailed discussion of these Regulations.)

The Personal Protective Equipment (EC Directive) Regulations 1992, as amended
Hearing protection devices in the workplace must be capable of reducing noise to the levels specified in these Regulations. Further all hearing protection devices must state the noise attenuation levels and the comfort level provided thereby. (See Chapter 7 for a detailed discussion of these Regulations.)

6

Fire safety

6.1. INTRODUCTION

Responsibility for fire safety in the UK is split between the local fire
authorities and the Health and Safety Executive. The fire authorities
are primarily responsible for issuing fire certificates and conducting
inspections, with the HSE responsible for most other matters, including the development of fire safety rules for construction and building
sites.

For these reasons, it is imperative for any person working in the
construction industry to be aware of the different fire safety regimes,
which are summarised below.

6.2. FIRE CERTIFICATION

The main source of information on fire certification in the UK is the
Fire Precautions Act 1971 and the subsidiary legislation passed
thereunder. Together these laws lay out the rules and requirements
for the issuance of fire certificates. These rules have been
promulgated by the Home Office, but are enforced by local authority

fire inspectors. To the extent that a construction project involves the restoration/refurbishment of an existing building, or involves the operation of a building on behalf of the owner, this aspect of the fire safety regime may be relevant to the construction industry.

6.2.1. The Fire Precautions Act 1971

At its most basic, the Fire Precautions Act 1971, as amended by the Fire Safety and Safety of Places of Sport Act 1987 and the Fire Precautions (Factories, Offices, Shops and Railway Premises) Order 1989 (collectively, the Fire Precautions Act) obliges occupiers of certain commercial premises to obtain a fire certificate for those premises.

To date the Home Office has specified only two types of premises that must have fire certificates, namely

- factories, offices and shops
- hotels and boarding houses.

Even so, many of these premises have been specifically excluded from the requirement for a fire certificate by the Fire Precautions (Factories, Offices, Shops and Railway Premises) Order 1989. They are excluded if

- not more than 20 people are at work in the premises at any one time, or
- not more than ten people are at work in the premises at any one time, on floors other than the ground floor.

These exceptions do not apply, however, to premises where substantial quantities of hazardous or highly flammable materials are stored. (For more details, see Section 6.6. on 'special premises'.)

6.2.2. Contents of fire certificate

To the extent that premises require a fire certificate, that certificate must specify

- the use or uses of the premises

- the means of escape from those premises in the event of fire
- how those means of escape can be used safely and effectively
- the location and type of alarms and fire warning systems for the premises
- the fire-fighting equipment available for use on the premises, and
- the particulars of any highly flammable materials or explosives used or stored on the premises.

In addition, the fire authority *may* require that the following information should be included in a fire certificate

- a method statement for maintaining the fire escape and fire-fighting equipment
- staff training on emergency procedures, and
- a restriction on the number of persons occupying the building.

Typically fire certificates are secured by the owner or occupier of a building after occupation. As such, it will often be the case that a contractor need not concern himself with fire certification. Having said that, however, the Fire Precautions Act requires that the owner or occupier notify the fire authorities whenever a building subject to a fire certificate is significantly extended or altered, to the extent that such alteration may necessitate a change to the fire certificate. Similarly, designers will need to be familiar with the requirements of the Act so as to ensure that the buildings they design are eligible for a fire certificate when complete.

6.3. SAFETY EQUIPMENT/FIRE PRECAUTIONS FOR EXEMPT PREMISES

Premises that are exempt from having a fire certificate must nevertheless maintain adequate fire safety equipment and precautions, in accordance with the Fire Precautions (Factories, Offices, Shops and Railway Premises) Order 1989.

Fire/smoke/radiation detectors
Fire alarms
Fire fighting appliances — manual (e.g. fire blankets,
fire extinguishers)
Fire fighting appliance — automatic (e.g. sprinklers,
closing of fire doors and automatic ventilation
systems)
Fire escapes and fire escape routes
Fire doors
Emergency exits
Emergency lighting
Fire drills and staff training
Fire fighting system inspections
Appointment of responsible persons
Notices of fire routes

*Figure 3. Example of fire safety equipment and
precautions*

Figure 3 provides a list of some fire safety equipment/precautions
that may be appropriate for premises like a construction site.
Obviously, every premises is unique and may require additional
precautions or equipment, as appropriate.

6.4. FIRE PRECAUTIONS (WORKPLACE) REGULATIONS 1997

On 1 December 1997, the Fire Precautions (Workplace) Regulations
1997 are due to come into force. When effective, these Regulations
will oblige owners and/or occupiers of premises to conduct fire assessments, with the goal of identifying all the fire risks inherent in those
premises. Following the assessment, that owner or occupier must take
all appropriate steps to limit the risks identified.

These regulations will apply to most places of work, with the exception of

- construction sites coming within the scope of the construction HSW Regulations (see Section 5.2.)
- premises already subject to a fire certificate pursuant to the Fire Precautions Act
- special premises (see Section 6.6.)
- mines/quarries
- sports grounds/stadia
- any movable structure, or
- certain offshore installations.

6.5. BUILDING REGULATIONS

Every building that undergoes construction, renovation, or alteration equivalent to a 'material change of use', must be put into compliance with the Building Regulations 1991, as amended by the Building Regulations (Amendment) Regulations 1997, including Part B of Schedule 1 to those Regulations in respect of fire safety. Given the fundamental nature of the Building Regulations, every contractor must ensure that its building satisfies the fire safety requirements before handing over the premises.

In particular, Part B of Schedule 1 provides that all buildings must

- be supplied with adequate fire escapes
- have fire resistant internal liners installed
- be designed to ensure structural stability for a reasonable period in the event of fire
- be designed to inhibit fire or smoke from concealed spaces
- have fire resistant walls and roof, and
- provide facilities for fire-fighters to gain access.

The Building Regulations are enforced by building inspectors appointed by the local authority, in conjunction with the fire authorities.

6.6. SPECIAL PREMISES

The HSE is the enforcing authority for fire regulations in respect of 'special premises', by virtue of the Fire Certificates (Special Premises) Regulations 1976. Premises are considered 'special' if they present an extraordinary risk of fire as a result of the presence of substantial quantities of hazardous or flammable materials and substances.

Table 10 contains a list of the substances which, if stored in the quantities stated, would necessitate obtaining a special premises certificate.

In addition, special premises certificates are required for the following premises

- explosives factories or magazines licensed under the Explosives Act 1875
- mines, quarries or nuclear installations
- premises, other than hospitals, at which particles can be charged up to 50 megavolts
- premises subject to the Ionising Radiations Regulation 1985, and
- with some limited exceptions, buildings constructed or used for temporary occupation during building operations or engineering construction works.

A special premises fire certificate will not be issued until the HSE is satisfied that the premises possess certain minimum fire safety standards, such as an adequate means of escape, adequate fire-fighting equipment and other fire precautions. No work may be undertaken on special premises until a fire certificate is granted by the HSE, and is on display at all times.

Related to these duties are those contained in the Dangerous Substances (Notification and Marking of Sites) Regulations 1990, which provide that a person in control of a site storing 25 tonnes or more of dangerous substances must notify the appropriate fire authority and the HSE. Furthermore, safety signs must be put into place identifying the fact that dangerous substances are located on that site.

Table 10. Materials requiring special premises certificates

Material/substances	Quantities
Liquefied petroleum gas, liquefied natural gas, liquefied flammable gas (consisting primarily of methyl acetylene) stored for purposes other than fuel	In excess of 100 tonnes per week
Highly flammable liquid stored under pressure	In excess of 50 tonnes per week (at its boiling point)
The manufacture of expanded cellular plastics	In excess of 50 tonnes per week
The manufacture or storage of liquid oxygen	In excess of 135 tonnes per week
Chlorine, except when stored for purification purposes	In excess of 50 tonnes per week
The manufacture or storage of ammonia or artificial fertilisers	In excess of 250 tonnes
The processing, manufacture, use or storage of phosgene	5 tonnes
The processing, manufacture, use or storage of ethylene oxide	20 tonnes
The processing, manufacture, use or storage of carbon disulphide	50 tonnes
The processing, manufacture, use or storage of acrylonitrile	50 tonnes
The processing, manufacture, use or storage of hydrogen cyanide	50 tonnes
The processing, manufacture, use or storage of ethylene	100 tonnes
The processing, manufacture, use or storage of propylene	100 tonnes
The processing, manufacture, use or storage of any flammable liquid not otherwise specified	400 tonnes

6.7. OTHER FIRE LEGISLATION

There are numerous statutory instruments relating to fire safety. The following is a list of some of those which may contain standards relevant to the construction industry.

- The Carriage of Dangerous Substances by Road Regulations 1996 specifies that if dangerous substances are transported to and from a construction site, the vehicle operator must ensure that the vehicle carries adequate fire-fighting equipment and that precautions are taken to prevent fire or explosions on that vehicle.
- The Borehole Sites and Operations Regulations 1995 requires borehole operators to prepare a health and safety document for the site, containing a plan for the prevention of fire and explosions. That plan should also lay out the fire precautions developed for the site, including measures for detecting and fighting fires.
- The colour, appearance, size and location of fire safety signs and signs identifying fire-fighting equipment are set out in the Health and Safety (Safety Signs and Signals) Regulations 1996. Examples of some of the fire precaution signs in use are shown in Figs 4, 5 and 6.

Figure 4. Fire Exit

Figure 5. Risk of Fire

Figure 6. Fire-fighting equipment

- The Highly Flammable Liquids and Liquefied Petroleum Gases Regulations 1972 require that flammable liquids be kept in specified containers and storage facilities and that adequate means of escape exist in the event of fire.

7

Personal protective equipment

7.1. INTRODUCTION

It has always been the HSE's position that personal protective equipment should be a 'last resort' for employers when selecting risk reduction measures. In other words, the HSE expects employers to establish measures that should not rely on a individual's willingness to use them. Having said that, it is clear that personal protective equipment is a fundamental part of most risk reduction programmes in the construction industry.

For that reason this Chapter considers this issue in some detail, beginning with a consideration of the general duties of employers in respect of personal protective equipment that are contained in the Personal Protective Equipment at Work Regulations, and then outlines certain legislation that deals with specific types of personal protective equipment.

Table 11 provides a list of some of the more common types of personal protective equipment.

Table 11. Personal protective equipment

Category of equipment	Equipment examples
Head protection	Crash helmets Cycling helmets Riding helmets Industrial safety helmets Scalp protectors
Eye protection	Safety spectacles Eye shields Goggles Welding filters Face-shields Hoods
Foot and leg protection	Safety boot or shoe Gaiters Foundry boots Wellington boots Anti-static footwear Conductive footwear
Hand and arm protection	Chain-mail gloves Leather gloves Thermal insulated gloves
Hearing protection	Ear muffs Ear plugs
Respiratory protective equipment	Dust respirators Respirators for gases, vapours and fumes Powered visors and helmets Breathing apparatus
Protective clothing and equipment	Coveralls/overalls Insulated clothing Chain-mail vests High visibility clothing Immersion suits Life-jackets

7.2. THE PERSONAL PROTECTIVE EQUIPMENT AT WORK REGULATIONS 1992

7.2.1. Assess need for PPE

The Personal Protective Equipment at Work Regulations 1992 (the PPE Regulations) along with its guidance, imposes duties on employers (including the self-employed) in respect of most personal protective equipment and clothing provided for use at the workplace.

In particular, the PPE Regulations oblige employers to make an assessment of their operations to determine when and where personal protective equipment is being provided for their employees. Once that determination has been made the employer must then assess whether he can reduce the need for personal protective equipment by reducing the risks of injury in another way, so far as is reasonably practicable.

While the PPE Regulations do not require employers to provide personal protective equipment to non-employees, employers must remember that they have duties under ss. 3 and 4 of the HSW Act to take steps to ensure the safety of any person who is not in their employ, but who may nevertheless be affected by their operation or who may be on their property. (See Chapter 2 for a discussion of those sections of the HSW Act.) For this reason an employer may deem it appropriate, for example, to provide hearing protection to a subcontractor's workers or to site visitors, even though the PPE Regulations do not specifically require it.

7.2.2. Provide effective PPE

To the extent that an employer determines that it is not appropriate to eliminate all personal protective equipment or clothing from the site, the PPE Regulations require that employer to ensure, so far as is reasonably practicable, that the equipment/clothing that is provided is the most effective to combat the risks faced by his employees. This is to be accomplished by the employer considering the suitability and compatibility of the equipment or the clothing, as well as the maintenance and accommodation provided therefor.

Equipment must be 'suitable'
When determining whether equipment is 'suitable', the employer must determine whether the equipment is

- appropriate for the risk(s) involved
- appropriate for the workplace conditions, which may require consideration of, for example, the physical effort required to perform the task, the methods of working, the length of time that the equipment must be worn, the requirements for visibility and communications
- ergonomically suited to the wearer or user
- capable of fitting the wearer correctly, and
- able to prevent or adequately limit the risk of injury, so far as is reasonable practicable.

When determining suitability, the Guidance Notes prepared for these Regulations suggest that it may be appropriate in some cases for the employer to

- consult with the employee(s) who will be using the equipment to get their views, and/or
- buy more than one size to ensure better comfort and fit.

All personal protective equipment used in the European Union must satisfy or comply with applicable standards, which are set out in the Personal Protective Equipment (EC Directive) Regulations 1992, as amended by the Personal Protective Equipment (EC Directive) (Amendment) Regulations 1993, the Personal Protective Equipment (EC Directive) (Amendment) Regulations 1994 and the Personal Protective Equipment (EC Directive) (Amendment) Regulations 1996. Collectively, these regulations set out the procedures for ensuring that personal protective equipment manufactured and supplied for use in the workplace complies with the essential safety requirements. This compliance is evidenced by affixing a 'CE' mark to that equipment. Suppliers of equipment are advised to refer to the guidance that accompanies these Regulations for details of those procedures.

Equipment must be 'compatible'
To the extent that an employee may face multiple hazards requiring more than one item of personal protective equipment/clothing, the PPE Regulations provide that employers must assess the compatibility of such equipment/clothing when used together. This assessment should limit the possibility that hazards are actually created by the use of more than one kind of personal protective equipment. For example, if both a helmet and respiratory equipment are required, the employer must assess whether it is possible to wear both pieces of equipment at the same time and still remain effective.

7.2.3. Maintain and repair PPE

Employers must ensure that the personal protective equipment that they provide is maintained in good working order. This maintenance would include appropriate cleaning, disinfection, examination, testing, replacement or repair. Ideally, maintenance should be conducted in accordance with a regular schedule and prescribed procedures.

Spare parts, as well as cleaning and disinfection facilities, should be made readily available for all reusable personal protective equipment/clothing. Alternatively, disposable equipment/clothing may be offered to employees, as appropriate.

7.2.4. Provide PPE accommodation

In addition to providing the equipment/clothing itself, the PPE Regulations oblige employers to provide suitable accommodation therefor. This accommodation might include, for example, lockers for storing overalls and boots, or cases for carrying transportable equipment such as safety spectacles or respiratory equipment. Further, to the extent that equipment/clothing becomes contaminated during use (e.g. by exposure to asbestos), it must be immediately segregated and decontaminated in an area away from uncontaminated equipment or clothing.

7.2.5. Provide information, instruction and training

The PPE Regulations also oblige employers to provide information, instruction and training to all persons using personal protective equipment/clothing. This duty stems from the more general duty in the HSW Act to provide information, instruction and training to employees.

The training provided by the employer should include information on

- the health and safety risks leading to the use of personal protective equipment
- the function of the equipment, in terms of reducing the risks of injury
- factors that could affect the performance of that equipment, such as working conditions (temperature, lighting, etc.), defects, compatibility with other equipment, etc.

In addition to this general information, employers must provide practical instruction on

- how to use, put on, wear and remove the equipment
- the procedures for maintaining the equipment and obtaining repairs or replacements, and
- the safe storage of the equipment.

All this information and training must be delivered in a manner that is comprehensible to the person receiving it. As such, if a person is deaf, instructions should be provided in writing, or if a person has a limited understanding of the English language, instructions should be provided in a language he can understand. Training of this type should be updated or refreshed as often as necessary.

7.2.6. Ensure use of equipment

The PPE Regulations specifically state that an employer must ensure that any personal protective equipment that is provided is, in fact, used properly. In addition, the employer must ensure that use of the equipment is appropriately supervised, and in accordance with manufacturers' suggestions, if available.

7.2.7. Employees' duties

In addition to the duties imposed on employers and the self-employed, the Regulations impose a number of duties on *employees*. Specifically, employees must

- use any equipment provided
- use the equipment in accordance with any instruction or training given
- take reasonable care of the equipment
- ensure that the equipment is returned to the accommodation provided, and
- report any obvious loss or defect, in accordance with the employer's reporting procedures.

No charge may be levied to an employee for personal protective equipment which is used only at work. This confirms s. 9 of the HSW Act which provides that:

> no employer shall levy or permit to be levied on any employee of his any charge in respect of anything done or provided in pursuance of any specific requirements of the relevant statutory provisions.

7.3. OTHER PERSONAL PROTECTIVE EQUIPMENT LEGISLATION

Not all personal protective equipment is regulated by the PPE Regulations. There are a number of statutes in place that set out standards for specific types of personal protective equipment. In the event of overlap, the more specific legislation will take priority over the PPE Regulations. Having said that, to the extent that an employer conducts an assessment of personal protective equipment/clothing that is sufficient in order to comply with the PPE Regulations, that assessment will also satisfy the requirements of the more specific legislation.

The following is a list of some of the specific personal protective equipment legislation that may affect a construction project, beginning with one of the most familiar and successful pieces of construction safety legislation, namely the Construction (Head Protection) Regulations 1989.

7.3.1. The Construction (Head Protection) Regulations 1989

The Construction (Head Protection) Regulations 1989 (the Head Protection Regulations) address the safety issues associated with head protection used on building sites and sites where engineering or construction works are being carried out.

In particular, the Head Protection Regulations require employers to assess the need for head protection on those sites. Head protection must be worn by everyone on site whenever there exists a foreseeable risk of injury to the head arising from causes other than the employee falling, such as objects falling on to them. In addition, employers *and* others who are in control of a site must ensure that all persons who work on that site (whether in their employ or not), wear their head protection whenever necessary. In this respect the Head Protection Regulations go beyond the PPE Regulations by imposing compliance duties on both employers and persons in control of premises, in recognition of the fact that construction sites are often shared by several employers.

The Head Protection Regulations also oblige every employee or self-employed person to wear the head protection supplied (with the exception of Sikhs who wear turbans), in conformance with any rules or directions prepared by the employer. Furthermore, an employee must report any loss or defect in the head protection to his employer as soon as possible.

Finally, the Head Protection Regulations set out the minimum safety standards for head protection. In particular, head protection must

- be of an appropriate size for the wearer
- have an adjustable headband, nape and chin strap

- comfortably accommodate thermal liners in cold weather
- have a flexible headband of adequate width, contoured both vertically and horizontally
- have a removable and absorbent sweat-band, which can be either replaced or cleaned
- have textile cradle straps, and
- have appropriate chin straps.

Finally, employers must ensure that any head protection supplied is maintained and repaired whenever necessary.

7.3.2. The Noise at Work Regulations 1989

These Regulations impose a number of duties on employers to provide suitable hearing protection when certain noise levels are reached. The suitability of this equipment must be assessed by the employer in accordance with the PPE Regulations, as appropriate. Obviously, any hearing protection provided by employers must be compatible with other personal protective equipment provided.

The HSE has produced a substantial amount of guidance on selecting appropriate hearing protection equipment, including Noise Guide No. 5, entitled *Types and Selection of Personal Ear Protectors.*

7.3.3. The Control of Asbestos at Work Regulations 1987

These Regulations oblige employers to provide adequate and suitable protective clothing and equipment to their employees who are exposed to asbestos. In addition, the employer must ensure that all protective clothing be properly segregated, labelled, cleaned or disposed of after exposure, as appropriate.

Employers also must provide appropriate respiratory equipment to employees who are exposed to asbestos at the workplace if it is not reasonable to prevent exposure in the first place. Again, that equipment must satisfy the general requirements of the PPE Regulations.

See Chapter 8 for details of the Control of Asbestos at Work Regulations.

7.3.4. The Control of Substances Hazardous to Health Regulations 1994

Regulation 7 of the Control of Substances Hazardous to Health Regulations 1994, as amended, (the COSHH Regulations) obliges all employers to provide appropriate respiratory protective equipment to any person who may be exposed to actionable levels of hazardous substances if that exposure can not be prevented.

Respiratory equipment is considered appropriate if it is

• suitable for the purpose, and
• complies with any applicable European Union standards in respect of design and/or manufacture.

The Regulations go on to provide that this equipment shall be regularly tested, examined, and maintained, as necessary. (See Chapter 8 for details of the COSHH Regulations.)

7.4. GUIDANCE ON PERSONAL PROTECTIVE EQUIPMENT

The HSE has produced a substantial amount of guidance on the various types of personal protective equipment, including, but not limited to, guidance on head protection, respiratory protective equipment, and others.

8

Hazardous substances

8.1. INTRODUCTION

Many types of work, and in particular construction work, require contact with or exposure to hazardous substances. Given this fact, and the fact that there are a vast number of such substances in existence, it is not surprising that there is a large amount of legislation in this area.

Accordingly, this Chapter will only attempt to highlight the hazardous substances legislation that are of particular concern to the construction industry, namely the Control of Substances Hazardous to Health Regulations, the Asbestos Regulations, the Ionising Radiations Regulations and the Explosives Regulations.

8.2. CONTROL OF SUBSTANCES HAZARDOUS TO HEALTH REGULATIONS

The Control of Substances Hazardous to Health Regulations 1994, as amended by the Control of Substances Hazardous to Health (Amendment) Regulations 1996 (collectively the COSHH

Regulations), are the widest-reaching of the hazardous substance regulations, both in terms of the number of substances they cover and in the scope of the duties that they impose on employers.

What follows is a summary of the COSHH Regulations, beginning with the definition of the term 'hazardous substances'.

8.2.1. Definition

The COSHH Regulations defines a 'hazardous substance' as any substance which

- is listed in the Approved Supply List of dangerous substances appended to the Chemicals (Hazard Information and Packaging for Supply) Regulations 1994, as amended by the Chemicals (Hazard Information and Packaging for Supply) (Amendment) Regulations 1996, and the Chemicals (Hazard Information and Packaging for Supply) (Amendment) Regulations 1997
- has a maximum exposure level or an occupational exposure standard, as listed in Schedule 1 to the COSHH Regulations
- is a biological agent
- represents a substantial concentration of airborne dust, or
- is any other substance that creates a comparable health hazard.

8.2.2. Employers' duties under COSHH

The main duties imposed on employers by the COSHH Regulations are

- to conduct an assessment of their operations to determine any risk of injury from exposure to hazardous substances, which assessment shall be repeated as often as necessary
- to prevent or minimise the likelihood of such exposure, so far as is reasonably practicable, by means other than the provision of personal protective equipment, such as
  - o total enclosure of all processes and systems generating hazardous substances
  - o reducing the quantity of hazardous substances generated
  - o reducing the quantity of hazardous substances put into use

- o limiting access to the areas where hazardous substances are used or kept, or to contaminated areas
- o prohibiting eating, drinking and smoking in areas where exposure is possible
- o providing suitable hygiene facilities, including washing and cleaning areas
- o safe storage, handling and disposal of such substances
- to ensure the proper use, maintenance and repair of all control measures and systems, so far as is reasonably practicable
- to provide adequate monitoring and health surveillance for all employees who may be exposed to hazardous substances, so far as is reasonably practicable, and
- to provide information, instruction and training to any person who may be exposed to hazardous substances.

If it is not reasonably practicable for an employer to limit the risk of exposure to hazardous substances by direct means (e.g. completed enclosure or elimination of that substance), the employer may provide suitable personal protective equipment as a means of risk reduction. Having said that, personal protective equipment is always to be considered the method of 'last resort' when reducing the risk of injury. (See Chapter 7 for details on personal protective equipment legislation.)

8.2.3. MEL and OES

As mentioned, many hazardous substances have been allocated a maximum exposure level (MEL) or an occupational exposure standard (OES) by the HSE. Those standards indicate the maximum quantity of a hazardous substance that a person may be exposed to, on either a short-term (MEL) basis or a long-term (OES) basis. In no event, may an employer permit exposure in excess of the levels listed in Guidance Note EH 40, entitled *Occupational Exposure Limits*.

The COSHH Regulations contain a number of other specific requirements in respect of hazardous substances. For more information, the reader is advised to refer to the Regulations, along with the ACoP and the Guidance Notes prepared for the Regulations.

8.3. ASBESTOS

Certainly exposure to asbestos represents one of the greatest health risks faced by construction workers today. This is largely due to the fact that asbestos materials were commonly used in older buildings that are now being refurbished or demolished.

In view of those hazards, a substantial amount of legislation has been put into place dealing specifically with asbestos, including the Control of Asbestos at Work Regulations 1987, the Control of Asbestos (Amendment) Regulations 1992, the Asbestos (Licensing) Regulations 1983 and the Asbestos (Prohibitions) Regulations 1992.

For purposes of these Regulations, 'asbestos' is generally defined to include the three most common categories of asbestos, namely

- crocidolite (blue asbestos)
- amosite (brown asbestos) and
- chrysotile (white asbestos).

The following is a brief description of that legislation.

8.3.1. The Control of Asbestos at Work Regulations 1987

The Control of Asbestos at Work Regulations 1987, as amended, are intended to limit the chance that an employee is exposed to asbestos. This is done by imposing a number of duties on employers, including

- identifying the likelihood of exposure to asbestos before commencing any work
- taking all appropriate steps to prevent exposure, or if prevention is not possible, to reduce the likelihood of exposure to the lowest levels reasonably practicable
- designating those parts of the workplace where persons could be exposed to asbestos in excess of the 'action levels' to be 'asbestos areas' (see below for definitions of those terms)
- designating those parts of the workplace where persons could be exposed to asbestos in excess of the 'control levels' to be 'respirator zones' (see below for definitions of those terms)

- providing adequate information, instruction and training to every employee who may be exposed to asbestos in excess of the action levels
- providing and maintaining suitable protective equipment and clothing, and providing suitable washing and cleaning facilities for all persons (see Chapter 7 for details on personal protective equipment requirements)
- providing adequate control systems to reduce the possibility of exposure to asbestos, which systems must be inspected and tested at regular intervals, and
- providing regular monitoring, health surveillance and medical examinations for any employee exposed to asbestos over the action levels, and keeping records of such activities.

Action levels and control levels

The Control of Asbestos (Amendment) Regulations 1992 state that if an employee may be exposed to asbestos in excess of 'action levels', then the employer must prevent that exposure. An action level is determined by reference to a person's cumulative maximum exposure to asbestos over a continuous twelve-week period. In addition, the employer must ensure that no employee is ever exposed to asbestos which exceeds the 'control limits', which are the maximum permitted exposure levels.

For precise details of these limits, the reader is advised to refer to the ACoP prepared for those Regulations, entitled *Work with Asbestos Insulation, Asbestos Coating and Asbestos Board.*

8.3.2. The Control of Asbestos (Amendment) Regulations 1992

The Control of Asbestos (Amendment) Regulations 1992 also require employers to prepare a 'plan of work' before commencing any demolition or construction work which could involve asbestos. The plan of work should lay out a suitable system of work for any work involving asbestos and must be made available to the HSE upon request.

8.3.3. The Asbestos (Licensing) Regulations 1983

Before commencing any asbestos removal operation, the Asbestos (Licensing) Regulations 1983 require employers to obtain an asbestos removal licence, unless

- the whole job will take two (2) hours or less to complete and no one will spend more than one (1) hour per week performing that work
- a self-employed person or an employer using his own employees is to do the work on his own premises, or
- the work consists solely of monitoring or collecting samples.

Licences are issued by the HSE, who has the power to impose such restrictions and limitations as it sees fit. In addition, notification must be sent to the enforcing authority not less than 28 days before any asbestos removal work is to commence. Further details on these Regulations is available from the HSE publication, A Guide to the Asbestos (Licensing) Regulations 1983.

8.3.4. The Asbestos (Prohibitions) Regulations 1992

The Asbestos (Prohibitions) Regulations 1992 prohibit certain activities involving asbestos, namely

- the supply/importation and some uses of amphibole asbestos
- the supply and certain uses of chrysotile asbestos, and
- asbestos spraying.

Recently, a proposal has been made to the European Commission for a complete ban on the use and supply of all types of asbestos, which has received qualified support in the UK.

8.4. IONISING RADIATION

8.4.1. The Ionising Radiations Regulations 1985

The Ionising Radiations Regulations 1985 were developed to limit the likelihood that an employee is exposed to ionising radiation at work. It does this by imposing a number of duties on employers and employees, and by establishing strict radiation dose limitation measures. 'Ionising radiations' include radiations which have the ability to shatter or ionise atoms, such as alpha and beta particles and gamma rays. These radiations are produced by, for example, X-ray and radiography equipment.

Ionising radiations are generally considered to be more dangerous than non-ionising radiations (e.g. ultraviolet light), which are currently not regulated by legislation. Having said that, however, a proposed European Directive on Physical Agents would restrict the use of non-ionising radiations in some circumstances. At the time of writing, it was not known if, or when, this Directive might be adopted.

Employers' duties
The Ionising Radiations Regulations provide that employers must

- not undertake ionising radiation work for the first time without notifying the HSE at least 28 days before commencement
- assess the risk that their employees could be exposed to ionising radiations, and to take all necessary steps to prevent or limit that exposure, so far as is reasonably practicable
- provide such control measures (e.g. shielding, ventilation, containment) and systems of work (e.g. the creation of controlled or supervised areas, the provision of personal protective equipment) as are necessary to limit the exposure levels, so far as is reasonably practicable
- prevent the exposure of persons to radiation in excess of the dose limits prescribed in the Schedules to the Regulations
- designate and demarcate 'supervised areas' and 'controlled areas', by reference to the possibility of radiation exposure in excess of permitted levels
- restrict access to controlled areas to 'classified persons', and only in accordance with an appropriate permit to work system
- appoint qualified radiation protection advisors to advise persons in respect of ionising radiation issues, and to notify the HSE of those appointments

- provide such information, instruction and training to all employees who work with ionising radiation, as necessary, and
- provide medical monitoring and health surveillance for all employees exposed to radiation, and to keep records of same.

Employees' duties

The Regulations provide that employees must

- not knowingly expose themselves or any other person to ionising radiations
- make full and proper use of any systems or equipment provided by their employer, and promptly report any defects in such systems/equipment, and
- not eat, drink or smoke, etc. in controlled areas.

Dose limitation measures

The Regulations also require employers to assess ionising radiation levels through the use of dosimetry services approved by the HSE. Records of those assessments must be kept by the employer. In addition, employers may need to notify the HSE in the event of over-exposure.

Further information on dose limitation and dosimetry services is available in the HSE Guidance entitled *Dose Limitation: Restriction of Exposure*.

8.4.2. Ionising Radiations (Outside Workers) Regulations 1993

The Ionising Radiations (Outside Workers) Regulations 1993 are of particular importance for the construction industry insofar as they impose duties on employers who employ labour from other European Union countries to work with ionising radiation.

Specifically, the Regulations require employers who supply contract labour to projects in other countries to assess the likelihood that those employees could be exposed to radiation while working on those projects. Following that assessment, the contracting employer may only permit workers to perform such outside work if they have been given suitable information and training on the risks associated with radiation exposure.

In addition, every outside worker must be given a radiation pass-book by the contracting employer, providing particulars about that worker, his medical condition and his current dose assessment. The dose limits in the radiation passbook must be updated, as appropriate, by the project employer.

More information on these requirements is available in the ACoP for these Regulations, entitled *Protection of Outside Workers Against Ionising Radiation Arising from any Work Activity*.

8.5. EXPLOSIVES

Explosives are another dangerous substance frequently encountered on a construction site. The Control of Explosives Regulations 1991 is the main body of legislation in this area.

Those Regulations oblige an employer to

- obtain a valid explosives certificate, verifying that he is a fit person to acquire and keep explosives
- keep no more than the stated quantity of explosives on site
- keep up-to-date records of explosive supplies, and
- report any lost explosives.

In addition, no explosives may be conveyed, kept or supplied, un-less they have been properly classified and labelled in accordance with the Classification and Labelling of Explosives Regulations 1983.

8.6. OTHER RELEVANT STATUTES

While the foregoing represent the main statutory instruments in respect of dangerous substances, there are many other instruments which may be relevant, including

- Notification of New Substances Regulations 1993

Notification of Installations Handling Hazardous Substances Regulations 1982
- Chemicals (Hazard Information and Packaging for Supply) Regulations 1994
- Chemicals (Hazard Information and Packaging for Supply) (Amendment) Regulations 1996
- Chemicals (Hazard Information and Packaging for Supply) (Amendment) Regulations 1997
- Carriage of Dangerous Goods (Classification, Packaging and Labelling) and Use of Transportable Pressure Receptacles Regulations 1996
- Carriage of Dangerous Goods by Road Regulations 1996
- Carriage of Dangerous Goods by Rail Regulations 1996
- Carriage of Dangerous Goods by Road (Driver Training) Regulations 1996
- Highly Flammable Liquids and Liquefied Petroleum Gases Regulations 1972
- Control of Lead at Work Regulations 1980

While this list should give you a good understanding of the law in respect of hazardous substances, it is by no means complete. Therefore, reference should be made to the HSE in the event of a specific query.

9

Accident reporting and investigation

9.1. INTRODUCTION

It is an unfortunate fact of life that accidents, incidents (i.e. 'near misses') and diseases follow on from working on construction sites. It is also an unfortunate fact that many of those lead to injury, fatalities and/or property damage. In view of these serious consequences, the Health and Safety Commission has required that many of these accidents, near misses, diseases and/or injuries be reported to the relevant health and safety enforcing authority. This is required by the provisions of the Reporting of Injuries, Diseases and Dangerous Occurrences Regulations 1995, which are outlined below.

In view of the fact that those reports often lead to investigations by enforcing authorities, this Chapter will also outline the extensive investigative powers of health and safety inspectors.

9.2. THE RIDDOR REGULATIONS

The primary focus for all accident reporting requirements is the Reporting of Injuries, Diseases and Dangerous Occurrences

Regulations 1995 (RIDDOR) which oblige employers and other responsible persons to report 'incidents', 'dangerous occurrences' and 'diseases' that occur at the workplace, including construction sites.

9.2.1. Definitions

In order to better understand the duties imposed by the RIDDOR Regulations, it is necessary to define a few key terms, including

- *Accidents*: although not defined in RIDDOR, this term has come to be defined in practice as an unexpected event at work (including acts of violence) which may cause injury. As such, both fatal accidents and accidents resulting in injury are to be reported under RIDDOR.
- *Dangerous occurrences*: Unlike the term 'accident', 'dangerous occurrences' have been defined in RIDDOR by reference to a list of particular incidents. Those incidents include, by way of example
  o the collapse of heavy equipment or scaffolding
  o the uncontrolled release of biological agents
  o the malfunction of breathing apparatus
  o an uncontrolled release from a pipe or pipeline, etc.
  In addition, an occurrence may be defined as 'dangerous' by reference to the fact that it occurs in a particular workplace. Examples of those dangerous occurrences include
  o in mines (e.g. the below ground ignition of gas, or the failure of breathing apparatus)
  o in quarries (e.g. the sinking of a craft, the unintended movement of slopes or faces, or plant explosions)
  o transport systems (e.g. accidents involving any kind of trains, accidents at level crossings, or serious transport congestion), or
  o offshore workplaces (e.g. unintentional hydrocarbon releases, fire or explosions on installations, subsidence or collapse of a sea bed, or falls into water of more than two metres).

- *Diseases*: This term refers to diseases that arise in conjunction with or, as a result of, a particular work activity or exposure to a particular work condition. Examples of diseases covered by RIDDOR include:
  - bursitis of the knee caused by physically demanding work leading to severe/prolonged friction or pressure;
  - pneumoconiosis caused by working with or handling fibrous materials such as asbestos, or
  - hand-arm vibration syndrome caused by working with vibrating machinery.

This is by no means a complete list of the accidents, dangerous occurrences and diseases that are covered by RIDDOR. For this reason, the reader is advised to refer to the Schedules appended to the Regulations and to the Guidance that has been produced by the HSE for these Regulations. A list of some of that Guidance is included in the Bibliography.

9.2.2. Reporting requirements

The Regulations impose a duty on the 'responsible person' for a site to submit a report to the relevant enforcing authority in the event of certain types of accidents, dangerous occurrences or diseases. For every workplace the enforcing authority will be either the HSE or the local authority. For construction sites, however, the enforcing authority is almost always the HSE.

The responsible person
The responsible person on a site generally will be either the employer or the person in control of that site at the time of the incident. On construction sites, for example, the responsible person might be the main contractor who has control over the site, or a subcontractor who employs a worker who is injured, or a subcontractor who supplies the equipment that is involved in a dangerous occurrence.

When to report

In particular, the responsible person must notify the enforcing authority every time

- a person dies on site from a work-related accident
- an employee suffers a major injury on site from a work-related accident
- a person suffers a major injury that is not work-related, but which is associated with a work activity (e.g. a customer or member of the public is injured while on the site)
- a person suffers a major injury as a result of an accident at a hospital
- any dangerous occurrence happens
- the written diagnosis of an occupational disease is received, or
- a person is incapacitated from work for three days, not including the day of injury.

How to report

Following a major injury, a fatal accident or a dangerous occurrence, the responsible person must notify the enforcing authority by the quickest possible means (usually by telephone). In most cases, this must be followed up with a written report on the appropriate version of Form 2508. In all other cases a written report must be sent, but there is no need for immediate notification. (Copies of the various versions of Form 2508 are included in Figure 7.)

If an employee is injured as a result of a work-related accident and later dies from those injuries, the responsible person must report the death in writing to the enforcing authority as soon as possible.

9.2.3. Record-keeping requirements

The responsible person for a site must also keep records of every incident, accident, dangerous occurrence, disease and/or injury that occurs at that workplace. Those records must be kept on that site for a minimum of three years and should include the following information

- the date and time of the event

Health and Safety at Work etc Act 1974
The Reporting of Injuries, Diseases and Dangerous Occurrences Regulations 1995

HSE
Health & Safety
Executive

Report of a case of disease

Filling in this form

This form must be filled in by an employer or other responsible person.

Part A

About you

1 What is your full name?

2 What is your job title?

3 What is your telephone number?

About your organisation

4 What is the name of your organisation?

5 What is its address and postcode?

6 Does the affected person usually work at this address?

Yes ☐ Go to question 7

No ☐ Where do they normally work?

7 What type of work does the organisation do?

Part B

About the affected person

1 What is their full name?

2 What is their date of birth?

/ /

3 What is their job title?

4 Are they

☐ male?

☐ female?

5 Is the affected person (tick one box)

☐ one of your employees?

☐ on a training scheme? Give details:

☐ on work experience?

☐ employed by someone else? Give details:

☐ other? Give details:

F2508A (01/96)

Continued overleaf

Figure 7 (pages 103–106). HSE Form 2508

Part C

The disease you are reporting

1 Please give:

* the name of the disease, and the type of work it is associated with; or

* the name and number of the disease *(from Schedule 3 of the Regulations – see the accompanying notes).*

2 What is the date of the statement of the doctor who first diagnosed or confirmed the disease?

/ /

3 What is the name and address of the doctor?

Continue your description here

Part D

Describing the work that led to the disease

Please describe any work done by the affected person which might have led to them getting the disease.

If the disease is thought to have been caused by exposure to an agent at work *(eg a specific chemical)* please say what that agent is.

Give any other information which is relevant.

Give your description here

Part E

Your signature

Signature

Date

/ /

Where to send the form

Please send it to the Enforcing Authority for the place where the affected person works. If you do not know the Enforcing Authority, send it to the nearest HSE office.

For official use	
Client number	Location number
Event number	☐ INV REP ☐ Y ☐ N

Figure 7. (continued)

HSE Health & Safety Executive

Health and Safety at Work etc Act 1974
The Reporting of Injuries, Diseases and Dangerous Occurrences Regulations 1995

Report of an injury or dangerous occurrence

Filling in this form
This form must be filled in by an employer or other responsible person.

Part A

About you
1 What is your full name?

2 What is your job title?

3 What is your telephone number?

About your organisation
4 What is the name of your organisation?

5 What is its address and postcode?

6 What type of work does the organisation do?

Part B

About the incident
1 On what date did the incident happen?

/   /

2 At what time did the incident happen?
(Please use the 24-hour clock eg 0

3 Did the incident happen at the above address?
Yes ☐ Go to question 4
No ☐ Where did the incident happen?
 ☐ elsewhere in your organisation – give the name, address and postcode
 ☐ at someone else's premises – give the name, address and postcode
 ☐ in a public place – give details of where it happened

If you do not know the postcode, what is the name of the local authority?

4 In which department, or where on the premises, did the incident happen?

F2508 (01/96)

Part C

About the injured person
If you are reporting a dangerous occurrence, go to Part F.
If more than one person was injured in the same incident, please attach the details asked for in Part C and Part D for each injured person.

1 What is their full name?

2 What is their home address and postcode?

3 What is their phone number?

4 How old are they?

5 Are they ☐ male? ☐ female?

6 What is their job title?

7 Was the injured person (tick only one box)
 ☐ one of your employees?
 ☐ on a training scheme? Give details:
 ☐ on work experience?
 ☐ employed by someone else? Give details of the employer:
 ☐ self-employed and at work?
 ☐ a member of the public?

Part D

About the injury
1 What was the injury? (eg fracture, laceration)

2 What part of the body was injured?

Continued overleaf

Figure 7. (continued)

3 Was the injury (tick the one box that applies)

☐ a fatality?

☐ a major injury or condition? (see accompanying notes)

☐ an injury to an employee or self-employed person which prevented them doing their normal work for more than 3 days?

☐ an injury to a member of the public which meant they had to be taken from the scene of the accident to a hospital for treatment?

4 Did the injured person (tick all the boxes that apply)

☐ become unconscious?

☐ need resuscitation?

☐ remain in hospital for more than 24 hours?

☐ none of the above.

Part E

About the kind of accident

Please tick the one box that best describes what happened, then go to Part G.

☐ Contact with moving machinery or material being machined

☐ Hit by a moving, flying or falling object

☐ Hit by a moving vehicle

☐ Hit something fixed or stationary

☐ Injured while handling, lifting or carrying

☐ Slipped, tripped or fell on the same level

☐ Fell from a height

How high was the fall?

[_____] metres

☐ Trapped by something collapsing

☐ Drowned or asphyxiated

☐ Exposed to, or in contact with, a harmful substance

☐ Exposed to fire

☐ Exposed to an explosion

☐ Contact with electricity or an electrical discharge

☐ Injured by an animal

☐ Physically assaulted by a person

☐ Another kind of accident (describe it in Part G)

Part F

Dangerous occurrences

Enter the number of the dangerous occurrence you are reporting. (The numbers are given in the Regulations and in the notes which accompany this form)

[_____]

Part G

Describing what happened

Give as much detail as you can. For instance

* the name of any substance involved
* the name and type of any machine involved
* the events that led to the incident
* the part played by any people.

If it was a personal injury, give details of what the person was doing. Describe any action that has since been taken to prevent a similar incident. Use a separate piece of paper if you need to.

Part H

Your signature

Signature

[_____]

Date

[__/__/__]

Where to send the form

Please send it to the Enforcing Authority for the place where it happened. If you do not know the Enforcing Authority, send it to the nearest HSE office.

For official use			
Client number	Location number	Event number	
[_____]	[_____]	[_____]	☐ INV REP ☐ Y ☐ N

Figure 7. (continued)

- the victim's details in the event of an injury or disease
- a brief description of the accident, disease or dangerous occurrence
- the date that the incident was first reported to the enforcing authority, and
- the method by which it was reported.

Security (Claims and Payments) Regulations
In addition to the record-keeping requirements imposed by RIDDOR, the Social Security (Claims and Payments) Regulations 1979 oblige all employers who operate a factory, or who employ more than ten employees, to keep an accident book on their premises.

Every injury that occurs on those premises must be recorded in that book, which must be kept for a minimum of three years after the last entry.

9.3. INVESTIGATING ACCIDENTS

Every accident or incident reported may be investigated by the relevant enforcing authority. In addition, enforcing authorities have the right to inspect any workplace within their jurisdiction at any time. In view of the breadth of this authority, it is necessary to consider briefly the investigative powers of health and safety enforcing authorities, which are found in a number of statutory instruments, and the HSW Act.

9.3.1. Executive's power to investigate

Section 14 of the HSW Act empowers the HSE to investigate any 'accident, occurrence, situation or other matter'. Furthermore, s. 20 of that Act gives health and safety enforcing authorities (i.e. the HSE or the local enforcing authority) extensive powers to conduct investigations whenever they have reason to believe an offence has occurred on a site, or that there may be a risk of serious injury thereon. Those powers include the right to

- enter onto premises for the purpose of conducting an investigation, whether accompanied by a constable or otherwise
- make such an examination as is necessary in the circumstances
- require that all or part of a premises remains undisturbed, as the inspector believes is necessary
- take such measurements, photographs, recordings and samples as the inspector believes are necessary
- cause any article or substance to be dismantled or tested if the inspector believes it is liable to cause danger
- take possession of any such article or substance, for as long as necessary
- require any person to answer any reasonable question that the inspector asks, and to hand over any information that the inspector reasonably requests
- require the production of, inspect and take copies of any books or documents that are relevant or which are necessary for the inspector to see, and
- require the provision of any facilities and assistance that the inspector requires.

In addition, the HSW Act also gives the inspector any other power which he may require for the purpose of his investigation.

Effectively, these powers entitle enforcing authorities to require the production of most documentation and to ask virtually any question, whether or not a person ultimately becomes a witness or even a defendant to a prosecution. The power to require information ends once the enforcing authority determines that it has sufficient evidence or information about a person to put them under a formal caution, as set out in the Police and Criminal Evidence Act 1984.

The person being investigated has a duty to co-operate with the enforcing authority and is expected to assist them with their enquiries as much as possible.

To the extent that an investigation of this type identifies health and safety violations or offences, the inspector has the power to initiate an enforcement proceeding. The reader is referred to Chapter 1 for a discussion of the various types of enforcement proceedings that may be brought.

9.3.2. *Safety representatives' power to investigate*

Safety representatives are appointed to represent the workforce in matters associated with health and safety at a workplace. These appointments are authorised by s. 2 of the HSW Act, in conjunction with the Safety Representatives and Safety Committees Regulations 1977, as amended by the Management of Health and Safety at Work Regulations 1992 and the Health and Safety (Consultation with Employees) Regulations 1996.

Those Regulations empower safety representatives to investigate any accident, dangerous occurrence, and every 'potential hazard'. As such, safety representatives can conduct investigations for the purposes of

- identifying causes of accidents
- investigating complaints by employees, and
- to consider the safety implications of new working methods and/or equipment.

The safety representative must be given access to all or any part of the employer's operations, provided that he has given the employer prior written notice before commencing his investigation. The safety representative may also inspect any documents and information that the employer is obliged to retain by statute upon prior notice to the employer, unless those records are confidential or contain health records. Finally, the employer is expected to provide the safety representative with reasonable facilities and assistance in this investigation.

10

First-aid on the site

10.1. INTRODUCTION

Section 2 of the HSW Act imposes a duty on all employers to provide for the health and safety of their employees. This broad duty has been interpreted, in part, as requiring employers to provide suitable first-aid facilities for their employees. In the UK, those requirements have been codified in the Health and Safety (First-Aid) Regulations 1981 (the First-Aid Regulations).

This Chapter will highlight the employer's duties to provide first-aid facilities, with particular emphasis given to those duties that are of relevance for the construction industry.

10.2. THE FIRST-AID REGULATIONS

10.2.1. Duty to provide first-aid equipment

The First-aid Regulations require employers to provide adequate first-aid equipment and facilities for all persons who may be injured while at a place of work. For this reason, employers must provide first-aid facilities for employees, as well as any other person who may be in the workplace, such as members of the public or subcontractors' employees.

10.2.2. Assessing first-aid requirements

As a first step, the First-Aid Regulations require every employer to assess the first-aid requirements of his workplace. As every workplace will have unique first-aid requirements, the ACoP for the Regulations advises employers to assess their operations to determine what facilities, equipment and first-aid personnel may be appropriate. When making this assessment, the ACoP suggests that employers consider a number of criteria, including

- the type of work that is being carried out at that workplace
- whether access to emergency treatment is difficult from that workplace
- whether employees regularly work away from the employer's own premises
- whether employees are peripatetic and if so, how often are they on the site, and
- whether more than one employers' employees are working together on that site.

Once the employer has assessed the requirements of his operation in respect of these criteria, he must provide the appropriate first-aid facilities. For some workplaces, this may mean offering a fully-equipped first-aid room, and in other situations a first-aid box will suffice. (See Section 10.3. for a discussion of those issues.)

The ACoP also suggests that employers should assess whether duplication of first-aid facilities can be avoided when several employers are working together on one site, such as a construction site. In those situations, the employers may be able to agree that the first-aid facilities provided by one person (e.g. the employer or the main contractor) will be made available to all workers, as appropriate.

10.2.3. Duty to provide first-aiders and/or appointed persons

Regulation 3(2) of the First-Aid Regulations obliges every employer to ensure that an adequate number of 'suitable persons' are available for rendering first-aid to any person injured on that site. 'Suitable persons' are defined to include both 'first-aiders' and 'appointed persons'. A 'first-aider' is any person who holds a current training certificate or any other person who has undergone training approved by the HSE. On the other hand, 'appointed persons' must have received sufficient first-aid training to be able to take charge in the event of an injury or illness on site in the absence of a first-aider, but do not need to have the same level of training as a first-aider.

The ACoP sets out the criteria for employers to consider when they are deciding how many first-aiders should be at a workplace, which includes

- the number of employees working on site
- the distribution of employees in the establishment
- the nature of the work
- size and location of the site
- whether or not there is shift work, and
- the distance between the site and medical services.

For low risk workplaces, such as an office building, the ACoP recommends that not less than one (1) first-aider be made available for every 50 employees. Alternatively, the employer may appoint at least one appointed person to be on site whenever employees are working.

Obviously, the number of first-aiders supplied will depend on the nature of the work being undertaken. For higher risk workplaces, such as construction sites, more first-aiders will be required, but in no event should there be less than one first-aider for every 50 employees.

Furthermore, if the employer's operations present specific or unusual hazards, the ACoP suggests that at least one of the first-aiders should receive specialised first-aid training in respect of those hazards. Examples of this might include training in the treatment of acid burns or cyanide poisoning.

10.2.4. Duty to inform

Employers must provide their employees with suitable and sufficient information on the first-aid arrangements that are available at that site. This information should include the location of the first-aid equipment, the location of the first-aid room (if applicable) and the names and locations of the first-aid personnel. The ACoP recommends that this information be provided to all employees at their initial induction training and that notices should be placed in conspicuous places throughout the workplace.

10.2.5. Duty applies to the self-employed

Every self-employed person must provide adequate first-aid equipment for his/her own use while working. That equipment should be appropriate for the hazards that are regularly faced by that person. Obviously, when several self-employed persons are working together on one site, (which is often the case on construction sites), one person may be able to assess and co-ordinate the first-aid facilities for everyone.

This is the type of safety co-ordination that is encouraged by the Management of Health and Safety at Work Regulations 1992, as amended, and the Construction (Design and Management) Regulations 1994. (See Chapters 1 and 3, respectively, for a discussion of those Regulations.)

10.2.6. Record keeping

Finally, the ACoP recommends that a record of all first-aid treatment should be kept by employers in a suitable location (e.g. in the first-aid room), which records should include

- the name of the person receiving treatment
- the person's occupation
- the date and time of the accident
- the circumstances of the accident
- details of the injury suffered, and

- details of the treatment given.

In addition, every place of work must maintain an accident book in compliance with the Reporting of Injuries, Diseases and Dangerous Occurrences Regulations 1995 and the Security (Claims and Payments) Regulations 1979. (See Chapter 9 for a discussion of those Regulations.)

10.3. FIRST-AID BOXES AND ROOMS

Every workplace should have at least one first-aid kit on site. The minimum contents of first-aid kits are suggested in the ACoP and are listed in Figure 8.

First-aid rooms are recommended by the ACoP for larger or hazardous workplaces, such as construction sites. If a first-aid room is deemed appropriate by the employer, a suitable person must be appointed to be responsible for that room and its contents. That person must be available at all times when employees are at work.

A card giving general first-aid guidance
Two (2) sterile eye pads, with attachment
Twenty (20) individually wrapped sterile adhesive dressings
Six (6) medium-sized individually wrapped sterile unmedicated wound dressings
Two (2) large individually wrapped sterile unmedicated wound dressings
Three (3) extra large individually wrapped unmedicated wound dressings
Six (6) individually wrapped triangular bandages
Six (6) safety pins
Sterile water or saline, if clean mains water is not available

Figure 8. Minimum contents of first-aid kit

115

Similarly, the first-aid room should be available at all times when employees are at work and should be used only as a first-aid room.

The room should be large enough to accommodate a couch or bed, and wide enough to accommodate a stretcher or a wheelchair. Again, the ACoP sets out the minimum contents of a first-aid room, which are listed in Figure 9.

Sink, with running hot and cold water
Drinking water, when clean water is not otherwise available
Soap
Paper towels
Smooth topped working surfaces
Suitable storage for first-aid materials
First-aid materials equivalent to those in a first-aid box
A couch with a waterproof surface with a clean pillow and blanket
Clean garments for use of first-aiders
A chair
A record book
A bowl

Figure 9. Minimum contents for first-aid room

10.4. EXCLUSIONS

There are a number of workplaces that are excluded from the application of the First-Aid Regulations. They include

- places of work where the Diving Operations at Work Regulations 1981 apply
- places of work where the Merchant Shipping (Medical Scales) (Fishing Vessels) Regulations 1974 apply

- places of work where the Merchant Shipping (Medical Scales) Regulations 1974 apply
- vessels which are registered outside the UK
- coal, iron, shale or fire clay mines, or
- places of work operated by the Crown.

11

Miscellaneous health and safety issues

11.1. INTRODUCTION

There are a number of health and safety hazards on a construction site that do not fit neatly into the preceding Chapters. For that reason they have been grouped together in this the Miscellaneous Chapter, which addresses

- manual handling
- display screen equipment, and
- safety signs.

11.2. MANUAL HANDLING

There is no doubt that construction work involves a number of manual labour activities such as digging, lifting, hammering, etc. Equally, there is little doubt that those activities can lead to injuries such as muscle strain, broken bones, sprains, etc. In fact, nearly two-thirds of all work-related accidents are caused by manual handling activities, with most of those occurring on construction sites.

In view of these statistics, a European Directive 90/269/EEC was adopted in 1990 that set out minimum safety standards for manual handling activities performed at work. To implement that Directive, the UK produced the Manual Handling Operations Regulations 1992 (the Manual Handling Regulations), which impose a number of duties on employers, as outlined below.

11.2.1. Duty to limit manual handling

The Manual Handling Regulations oblige every employer (and self-employed person) to take all reasonably practicable steps to limit the need for employees to perform manual handling operations while at work. 'Manual handling operations' are those that require transporting or supporting any load by hand or through the use of bodily force (e.g. lifting, putting down, pushing, pulling, carrying or moving). This definition is broad enough to include activities that are accomplished solely through human effort, as well as activities that may involve the use of mechanical assistance (e.g. a hoist), to the extent that such activity involves physical effort.

The employer is expected to conduct an assessment of his operations to determine where manual handling activities are being performed and the likelihood that those activities could lead to injury. Once those risks have been identified, the employer is then obliged to eliminate the need for manual handling, so far as is reasonably practicable.

11.2.2. Duty to assess manual handling tasks remaining

To the extent that an employer is unable to eliminate manual handling activities, he is expected to make a second assessment of the remaining manual handling activities to identify the continuing risks to safety. That assessment should take into consideration the factors set out in Schedule 1 to those Regulations, which are listed in Table 12.

Table 12. Factors to consider when assessing risk associated with manual handling operations

Factors	Considerations
Does the *task* involve	holding or manipulating loads at distances from trunk?
	awkward bodily movement or postures, including twisting, stooping or reaching upwards?
	excessive movement of loads, especially excessive lifting or lowering, carrying over long distances?
	excessive pushing or pulling of loads?
	sudden movement of loads?
	frequent or prolonged physical effort?
	insufficient rest or recovery from effort?
	a rate of work that cannot be changed?
Are the *loads*	heavy?
	bulky or unwieldy?
	difficult to grasp?
	unstable or with contents that are likely to shift?
	sharp, hot or otherwise potentially damaging?
Does the *working environment* have	space constraints preventing good posture?
	uneven, slippery or unstable floors?
	variation in floor levels or work surfaces?
	extremes of temperatures or humidity?
	conditions causing ventilation problems or gusts or wind?
	poor lighting?
Does the *individual have the capability* to perform tasks which	require unusual strength, height, etc.?
	present a danger to pregnant women or a person who has health problems?
	require special training and/or information?
Other factors	is the person's movement hindered by personal protective equipment and/or clothing?

Once the risks associated with the remaining manual handling tasks have been identified, the employer must take all appropriate steps, so far as is reasonably practicable, to reduce those risks. Those steps might include

- the provision of mechanical aids for completing the task ˇ
- a change in working methods to reduce the likelihood of injury associated with the task (e.g. purchasing smaller boxes of merchandise so as to make them easier to move/unload), or
- the provision of information on each load, including the weight, the location of its heaviest side and its centre of gravity.

These assessments must be reviewed and revised by the employer whenever he has reason to believe that they are no longer valid or that there has been a significant change in the nature of the manual handling operations at that workplace.

11.2.3. Employees' duty

Every employee who is supplied with a system of work in respect of manual handling must make full and proper use of that system. This duty is an extension of the general duty imposed by the Management of Health and Safety at Work Regulations 1992 as amended, which obliges employees to make full and proper use of the equipment provided by their employers. (See Section 1.3. for a detailed discussion of those Regulations.)

11.2.4. Guidance on Manual Handling

All of these duties are considered in great detail in the HSE's Guidance on the Manual Handling Regulations. In particular, the Guidance offers advice to employers on assessing their manual handling operations and methods for reducing risk. The Guidance also provides a sample manual handling assessment checklist for use by employers, which is reproduced in Fig. 10.

Example of an assessment checklist

Manual handling of loads

EXAMPLE OF AN ASSESSMENT CHECKLIST

Note: This checklist may be copied freely. It will remind you of the main points to think about while you:
- consider the risk of injury from manual handling operations
- identify steps that can remove or reduce the risk
- decide your priorities for action.

SUMMARY OF ASSESSMENT	Overall priority for remedial action: Nil / Low / Med / High*
Operations covered by this assessment:	Remedial action to be taken: ....
Locations:	Date by which action is to be taken: ....
Personnel involved:	Date for reassessment: ....
Date of assessment:	Assessor's name: .... Signature: ....

*circle as appropriate

Section A - Preliminary:
Q1   Do the operations involve a significant risk of injury?   Yes / No*
     If 'Yes' go to Q2. If 'No' the assessment need go no further.
     If in doubt answer 'Yes'. You may find the guidelines in Appendix 1 helpful.
Q2   Can the operations be avoided / mechanised / automated at reasonable cost?   Yes / No*
     If 'No' go to Q3. If 'Yes' proceed and then check that the result is satisfactory.
Q3   Are the operations clearly within the guidelines in Appendix 1?   Yes / No*
     If 'No' go to Section B. If 'Yes' you may go straight to Section C if you wish.

Section C - Overall assessment of risk:
Q   What is your overall assessment of the risk of injury?   Insignificant / Low / Med / High*
     If not 'Insignificant' go to Section D. If 'Insignificant' the assessment need go no further.

Section D - Remedial action:
Q   What remedial steps should be taken, in order of priority?
   i ....
   ii ....
   iii ....
   iv ....
   v ....

And finally:
     - complete the SUMMARY above
     - compare it with your other manual handling assessments
     - decide your priorities for action
     - TAKE ACTION.................AND CHECK THAT IT HAS THE DESIRED EFFECT

Figure 10 (above and overleaf). Manual handling checklist

123

Section B - More detailed assessment, where necessary:

Questions to consider: (If the answer to a question is 'Yes' place a tick against it and then consider the level of risk)	Yes	Level of risk: (Tick as appropriate) Low Med High	Possible remedial action: (Make rough notes in this column in preparation for completing Section D)
The tasks - do they involve: ◆ holding loads away from trunk? ◆ twisting? ◆ stooping? ◆ reaching upwards? ◆ large vertical movement? ◆ long carrying distances? ◆ strenuous pushing or pulling? ◆ unpredictable movement of loads? ◆ repetitive handling? ◆ insufficient rest or recovery? ◆ a workrate imposed by a process?			
The loads - are they: ◆ heavy? ◆ bulky/unwieldy? ◆ difficult to grasp? ◆ unstable/unpredictable? ◆ intrinsically harmful (eg sharp/hot?)			
The working environment - are there: ◆ constraints on posture? ◆ poor floors? ◆ variations in levels? ◆ hot/cold/humid conditions? ◆ strong air movements? ◆ poor lighting conditions?			
Individual capability - does the job: ◆ require unusual capability? ◆ hazard those with a health problem? ◆ hazard those who are pregnant? ◆ call for special information/training?			
Other factors - Is movement or posture hindered by clothing or personal protective equipment?			

Deciding the level of risk will inevitably call for judgement. The guidelines in Appendix 1 may provide a useful yardstick. **When you have completed Section B go to Section C.**

Figure 10. (continued)

11.3. DISPLAY SCREEN EQUIPMENT

The use of display screen equipment (e.g. personal computers, micro-fiche viewers) is an inevitable part of any industry these days — including the construction industry — whether in the office or as part of the engineering design function.

An unfortunate consequence of this dramatic increase in use is an increase in the number of injuries associated with such equipment, including

- musculoskeletal disorders
- visual fatigue, and
- stress.

For these reasons, European Directive 90/270/EEC was adopted, which set out minimum safety standards for the use of display screen equipment. That Directive was implemented in the UK by the Health and Safety (Display Screen Equipment) Regulations 1992 (the Display Screen Regulations), which set out a number of duties for employers, as well as laying out minimum standards for the equipment itself.

11.3.1. Definitions

The Display Screen Regulations impose duties in respect of 'display screen equipment'. That term is defined in the Guidance to the Regulations to include all screened equipment, ranging from personal computers, liquid crystal displays, to cathode ray tube screens.

Certain display screen equipment is excluded from these Regulations, including equipment that is

- used in control cabs for vehicles/machinery
- used on board a means of transportation
- intended for public operation
- a portable system not in prolonged use
- used in calculators, cash registers, window typewriters, ⟨
- used mainly to show television or films.

For this reason most electronic equipment with a display screen that is used in an office probably will be covered by these Regulations, provided that it is used by either a 'user' or an 'operator'. A 'user' is an *employee* who habitually uses this equipment as a normal part of his work. On the other hand, an 'operator' is a *self-employed person* who habitually uses this equipment as a normal part of his work. The Guidance to the Display Screen Regulations suggests that a person uses their equipment 'habitually' whenever

- his/her job depends on the use of display screen equipment
- the worker has no discretion whether or not to use that equipment
- significant training or skills is required to use that equipment
- the worker uses the equipment for more than one hour at a time, several times a day
- the job depends on the fast transfer of data between the worker and the screen, and
- the job demand high levels of concentration and attention to avoid error.

11.3.2. Duty to assess workstations

The Display Screen Regulations also consider display equipment in its wider context, i.e. as part of a 'workstation'. A workstation includes the desk, the work chair, printer, modem, as well as the immediate environment (i.e. the lighting, temperature, noise, space, etc.) around that station. The component parts to a workstation are listed in Table 13.

Employers must analyse every workstation provided to users and to operators, to assess both the health and safety hazards that may be associated with the use of that workstation and the likelihood that a person could be injured by those hazards. That assessment should be reviewed whenever the employer has reason to believe that it is no longer valid or there has been a significant change in the workstation.

To the extent that risks are identified in that assessment, the employer must reduce those risks to the lowest extent reasonably practicable. This may be accomplished by adjusting the existing workstation equipment (e.g. raising the height of the screen, or

Table 13 (below and overleaf). Minimum requirements for display screen workstations

Workstation components	Minimum standards
Display screen	On-screen characters that are well-defined and of adequate size No flickering and a stable image Adjustable brightness/contrast controls Screen that swivels and tilts Screen free of glare/reflections
Keyboard	Keyboard that tilts and which is separate from the screen Sufficient space in front of keyboard for arms to rest Matt keyboard surface Characters arranged so as to facilitate use Characters legible
Work desk/surface	Sufficiently large working surface, with non-reflective surface Stable document holder Adequate space to permit a comfortable position
Work chair	Stable chair which allows freedom of movement and a comfortable position Adjustable seat height Seat back adjustable in height and tilt Footrest, as appropriate
Space requirement	Sufficient space provided for freedom of movement
Lighting	Lighting is satisfactory to ensure appropriate contrast between screen and background Glare or reflections prevented by redirected lighting
Reflections and glare	Equipment located so as to avoid direct glare or reflections Windows fitted with adjustable coverings, as appropriate

Table 13. (continued)

Workstation components	Minimum standards
Noise	Noise levels which do not distract or disrupt speech
Heat	Excessive heat is prevented
Radiation	Radiation reduced to negligible levels, if an issue
Humidity	Adequate humidity levels
Interface between computer and user/operator	Software appropriate for the task, easy to use and adaptable
	Systems that provide feedback on performance
	Information displayed at appropriate speed
	Software ergonomics principles are applied

lowering the work chair), or by providing new workstation equipment (e.g. a footrest or an anti-reflective screen), as appropriate. The Guidance to the Regulations provides a thorough discussion of the options available to employers for reducing risks.

In addition, the component parts of every workstation must meet the minimum standards set out in the Schedules to the Regulations, which are as listed in Table 13. Having said that, the Guidance to the Regulations suggest that full compliance with BS EN 29241 (Parts 1–3) will normally be sufficient to comply with the requirements of the Schedule.

11.3.3. Eye and eyesight examinations

Upon request, employers must provide free eye and eyesight examinations to all employees who are habitual *users* (but not operators) of display screen equipment, not less than once every ten years, or sooner if an employee experiences vision problems related to the use of their equipment. Normal corrective devices (e.g. spectacles) must be supplied to any employee who specifically requires them for use of their display screen equipment. The employers' duty to provide eye wear extends only to basic spectacles and frames, with any additional requirements to be at the employee's own cost.

11.3.4. Information and training
Employers must provide users and operators of display screen equipment with adequate information on the health and safety risks associated with using the workstation and on the steps taken by the employer to limit those risks. In addition, employers must provide training to *users* (but not operators) on the correct method for using the workstation. This information and training must be updated by the employer whenever appropriate.

11.4. SAFETY SIGNS/SIGNALS

Safety signs and signals are an essential part of every employer's health and safety management system. This is particularly true in workplaces such as construction sites, where the hazards are numerous and not always obvious. For this reason it is fundamental that safety signs and signals be easily seen and immediately understood by everyone who visits a workplace, including for example, non-English speaking workers or visitors.

To accommodate this need, the safety sign legislation in the UK recently underwent a significant reworking to put it in line with European Council Directive 92/58/EEC. Those new requirements are contained in the Health and Safety (Safety Signs and Signals) Regulations 1996 (the Safety Signs Regulations), which set out the minimum requirements for safety signs at work, and impose a number of duties on both employers and others in control of premises.

11.4.1. Safety signs requirements
Every employer must conduct an assessment of the risks inherent in his operations and take steps to limit the risks identified, so far as is reasonably practicable, to satisfy the requirements of the Management of Health and Safety at Work Regulations 1992, as amended (see Chapter 1 for a discussion of those Regulations). To the extent that a risk has been identified which cannot be controlled by direct

means, the Safety Signs Regulations permit employers to provide safety signs for purposes of reducing risk. In this way, the use of safety signs is to be considered a 'last resort' method for an employer's risk reduction programme.

Employers must always use the correct sign/signal for the hazard or condition identified. Typically, safety signs serve one of four possible purposes, namely

- to prohibit action from being taken (e.g. unauthorised entry is prohibited)
- to make certain actions mandatory (e.g. hard hats must be worn)
- to provide warning of a hazard (e.g. danger of electric shock), or
- to announce a non-dangerous place or place of safety (e.g. f ᴧ escape route).

Schedule 1 to the Regulations sets out the shape, size, colour and location of signs/signals to ensure that uniformity. In this way

- red always signifies prohibited actions
- blue always signifies mandatory actions
- yellow always signifies hazards or danger, and
- green always signifies non-dangerous areas.

Every person coming on to the site should be able to identify whether a hazard exists or whether a certain action is required or prohibited — whether or not they read English.

Schedule 1 also sets out the minimum standards for other types of signs, signals or warnings, as follows

- signboards, containers and pipes (part III)
- fire fighting equipment (part IV)
- signs used for traffic routes, dangerous locations and obstacles (part V)
- illuminated signs (part VI)
- acoustic signals (part VII)
- verbal communications (part VIII), and
- hand signals (part IX).

Examples of some safety signs are provided in Figs 11, 12 and 13.

Figure 11. Fire Exit — Left

Figure 12. Hard Hats Required

Figure 13. Access Prohibited

11.4.2. Other duties

The Safety Signs Regulations also oblige employers to

- ensure that all safety signs and warning equipment are properly maintained
- provide their employees with suitable information and training as to the meaning of those signs/warning, and
- describe the correct action to take in response to the signs or warnings.

Table of statutes

Companies Directors Disqualification Act 1986
Employers' Liability (Compulsory Insurance) Act 1969
Explosives Act 1875
Factories Act 1961
Fire Precautions Act 1971
Fire Safety and Safety of Places of Sport Act 1987
Health and Safety at Work etc. Act 1974
Occupiers' Liability Act 1957
Occupiers' Liability Act 1984
Offices, Shops and Railway Premises Act 1963
Police and Criminal Evidence Act 1984

Table of statutory instruments and orders

Asbestos (Licensing) Regulations 1983 (SI 1983/1649)
Asbestos (Prohibitions) Regulations 1992 (SI 1992/3067)
Borehole Sites and Operations Regulations 1995 (SI 1995/2038)
Building Regulations 1991 (SI 1991/2768)
Building Regulations (Amendment) Regulations 1997 (SI 1997/1904)
Carriage of Dangerous Goods by Road (Driver Training) Regulations 1996 (SI 1996/2094)
Carriage of Dangerous Goods by Road Regulations 1996 (SI 1996/2095)
Carriage of Dangerous Goods by Rail Regulations 1996 (SI 1996/2089)
Carriage of Dangerous Goods (Classification, Packaging and Labelling) and Use of Transportable Pressure Receptacles Regulations 1996 (SI 1996/2092)
Chemicals (Hazard Information and Packaging for Supply) Regulations 1994 (SI 1994/3247)
Chemicals (Hazard Information and Packaging for Supply) (Amendment) Regulations 1996 (SI 1996/1092)
Chemicals (Hazard Information and Packaging for Supply) (Amendment) Regulations 1997 (SI 1997/1460)
Classification and Labelling of Explosives Regulations 1983 (SI 1983/1140)
Construction (Design and Management) Regulations 1994 (SI 1994/3140)
Construction (General Provisions) Regulations 1961 (SI 1961/1580)
Construction (Head Protection) Regulations 1989 (SI 1989/2209)
Construction (Health, Safety and Welfare) Regulations 1996 (SI 1996/1592)
Construction (Health and Welfare) Regulations 1966 (SI 1966/95)

135

Construction (Lifting Operations) Regulations 1961 (SI 1961/1581)
Construction (Working Places) Regulations 1966 (SI 1966/94)
Control of Asbestos at Work Regulations 1987 (SI 1987/2115)
Control of Asbestos (Amendment) Regulations 1992 (SI 1992/3068)
Control of Explosives Regulations 1991 (SI 1991/1531)
Control of Lead at Work Regulations 1980 (SI 1980/1248)
Control of Substances Hazardous to Health Regulations 1994 (SI 1994/3246)
Control of Substances Hazardous to Health (Amendment) Regulations 1996 (SI 1996/3138)
Dangerous Substances (Notification and Marking of Sites) Regulations 1990 (SI 1990/304)
Draft Fire Precautions (Places of Work) Regulations
Draft Lifts Regulations (available from the Department of Transportation)
Electricity at Work Regulations 1989 (SI 1989/635)
Fire Certificates (Special Premises) Regulations 1976 (SI 1976/2003)
Fire Precautions (Factories, Offices, Shops and Railway Premises) Order 1989 (SI 1989/76)
Fire Precautions (Workplace) Regulations 1997 (SI 1997/1840)
Health and Safety (Consultation with Employees) Regulations 1996 (SI 1996/1513)
Health and Safety (Display Screen Equipment) Regulations 1992 (SI 1992/2792)
Health and Safety (First-Aid) Regulations 1981 (SI 1981/917)
Health and Safety Information for Employers Regulations 1989 (SI 1989/682)
Health and Safety Information for Employees (Modifications and Repeals) Regulations 1995 (SI 1995/2923)
Health and Safety (Safety Signs and Signals) Regulations 1996 (SI 1996/341)
Highly Flammable Liquids and Liquefied Petroleum Gases Regulations 1972 (SI 1972/917)
Hoists Exemption Order 1962 (SI 1962/715)
Hoists Exemption (Amendment) Order 1967 (SI 1967/759)
Ionising Radiations Regulations 1985 (SI 1985/1333)
Ionising Radiations (Outside Workers) Regulations 1993 (SI 1993/2379)
Lifting Plant and Equipment (Record of Test and Examination, etc.) Regulations 1992 (SI 1992/195)
Lifts Regulations 1997 (SI 1997/831)

Management of Health and Safety at Work Regulations 1992 (SI 1992/2051)

Management of Health and Safety At Work (Amendment) Regulations 1994 (SI 1994/2865)

Manual Handling Operations Regulations 1992 (SI 1992/2793)

Noise at Work Regulations 1989 (SI 1989/1790)

Notification of Installations Handling Hazardous Substances Regulations 1982 (SI 1982/1357)

Notification of New Substances Regulations 1993 (SI 1993/3050)

Personal Protective Equipment (EC Directive) Regulations 1992 (SI 1992/3139)

Personal Protective Equipment (EC Directive) (Amendment) Regulations 1993 (SI 1993/3074)

Personal Protective Equipment at Work Regulations 1992 (SI 1992/2966)

Personal Protective Equipment (EC Directive) (Amendment) Regulations 1994 (SI 1994/2326)

Personal Protective Equipment (EC Directive) (Amendment) Regulations 1996 (SI 1996/3039)

Provision and Use of Work Equipment Regulations 1992 (SI 1992/2932)

Reporting of Injuries, Diseases and Dangerous Occurrences Regulations 1995 (SI 1995/3163)

Safety Representatives and Safety Committees Regulations 1977 (SI 1977/500)

Security (Claims and Payments) Regulations 1979 (SI 1979/628)

Social Security (Claims and Payments) Regulations 1979 (SI 1979/628)

Supply of Machinery (Safety) Regulations 1992 (SI 1992/3073)

Supply of Machinery (Safety) (Amendment) Regulations 1994 (SI 1194/2063)

Workplace (Health, Safety and Welfare) Regulations 1992 (SI 1992/3004)

Bibliography

Access to Tower Cranes (PM 9)

Approved Code of Practice: Control of Substances Hazardous to Health Regulations 1994 (L 5)

Approved Code of Practice: First-Aid at Work (ACoP 42) (rev.)

Approved Code of Practice: Management of Health and Safety at Work Regulations 1992 (L 21)

Approved Code of Practice: Managing Construction for Health and Safety — The Construction (Design and Management) Regulations 1994 (L 54)

Approved Code of Practice: The Control of Asbestos at Work 1987 (L 26)

Approved Code of Practice: Work with Asbestos Insulation, Asbestos Coating and Asbestos Board (L 28)

Asbestos: Exposure Limits and Measurement of Airborne Dust and Concentrations (EH 10)

A Brief Guide to the Requirements for the Control Noise at Work 1992 (IND (G) 75 L)

British Standards BS EN 29241 (Parts 1–3) On Display Screen Equipment

Code Of Practice for Fire Precautions in Factories, Offices, Shops and Railway Premises Not Required to Have a Fire Certificate)

Designing for Health and Safety in Construction: A Guide for Designers on the Construction (Design and Management) Regulations 1994

Display Screen Equipment: Guidance on the Regulations (L 26)

Dose Limitation: Restriction of Exposure (L 7)

Electrical Safety on Construction Sites (HS (G) 141)

Enclosures Provided for Work with Asbestos Insulation, Coatings and Insulation Board (EH 51)

Everyone's Guide to RIDDOR '95 (HSE 31)

Excavators Used as Cranes (PM 42)

139

Eye and Face Protection (CIS 31)
Fire Precautions Act 1971: Fire Safety at Work
First-Aid at Work: General Guidance for Inclusion in First-Aid Boxes (IND (G) 4P)
First-Aid Needs in Your Workplace: Your Questions Answered. (IND (G) 4)
General Access Scaffolds (CIS 3)
General and Specialist Clothing (CIS 33)
Getting to Grips with Manual Handling: A Short Guide for Employers (IND (L) 93)
Gloves (CIS 34)
Grin and Wear It (IND (G) 137)
Guidance Note: Occupational Exposure Limits (EH 40)
Guidance Notes on the Supply of Machinery (Safety) Regulations
A Guide to the Asbestos (Licensing) Regulations 1983 (L 11)
A Guide to the Control of Explosives Regulation 1991 (L 10)
A Guide to the Health and Safety at Work etc. Act 1974 (L 1)
A Guide to the Lifting Plant and Equipment (Record of Test and Examination, etc.) Regulations (L 20)
A Guide to Managing Health and Safety in Construction
A Guide to the Packaging of Explosive for Carriage Regulations (L 13)
A Guide to the Reporting of Injuries, Diseases and Dangerous Occurrences Regulations 1995 (L 73)
Head Protection (CIS 29)
Hearing Protection (CIS 30)
Inclined Hoists Used in Building and Construction Work (PM 63)
Joint Code of Practice on the Protection from Fire on Construction Sites and Buildings Undergoing Renovation
Keep Your Top On (IND (G) 147)
Law Commission Paper 127: 'Legislating the Criminal Code: Involuntary Manslaughter'
Lifts, Thorough Examinations and Testing (PM 7)
Lighting at Work (HS (G) 38)
Manual Handling: Guidance on the Regulations (L 23)
Manual Handling: Lighten the Load (IND (G) 146L)
Manual Handling: Solutions You Can Handle (HS (G) 115)
Noise Assessment, Information and Controls on Noise Guides 3 to 8 (HS (G) 56)

Noise in Construction: Further Guidance to the Noise at Work Regulations 1989 (IND (G) 127L)

Passenger Carrying Paternosters (PM 8)

Personal Protective Equipment (PPE): Principles, Duties and Responsibilities (CIS 28)

Personal Protective Equipment at Work Regulations 1992: Guidance on the Regulations (L 25)

The Problems of Asbestos Removal at High Temperatures (EH 57)

Proposals for the Implementation of the Lifting Aspects of the Amending Directive to the Use of Work Equipment Directive (CD 116)

The Protection of Persons Against Ionising Radiations Arising from any Work Activity: The Radiations Regulations: Approved Code of Practice (L 58)

Protection of Outside Workers Against Ionising Radiations: The Ionising Radiations (Outside Workers) Regulations 1993 (L 49)

The Provision, Use and Maintenance of Hygiene Facilities for Work with Asbestos Insulation and Coatings. (EH 47)

Respiratory Protective Equipment: A Practical Guide for Users (HS (G) 53)

Respiratory Protective Equipment (CIS 32)

The Robens Report (1972)

Safe Erection of Structures: Part 3 (GS 28/3)

Safe Use of Ladders, Step Ladders and Trestles (GS 31)

Safe Working with Overhead Cranes (PM 55)

Safety Footwear (CIS 35)

Safety of Power-Operated Mast Work Platforms (HS (G) 23)

Safety Signs and Signals: Guidance on the Regulations (L 64)

A Short Guide to Personal Protective Equipment at Work (IND (G) 174)

Signpost to the Health and Safety (Safety Signs and Signals) Regulation 1996 (IND (G) 184L)

A Step By Step Guide to COSHH Assessments (HS (G) 97)

Suspended Access Equipment (PM 30)

Temporarily Suspended Access Cradles and Platforms (CIS 5)

Tower Scaffolds (CIS 10)

Tower Scaffolds (GS 42)

The Training of Crane Drivers and Slingers (GS 39)

Training Operatives and Supervisors for Work with Asbestos Insulation and Coatings. (EH 50)

VDUs: An Easy Guide to the Regulations. How to Comply with the Health and Safety (Display Screen Equipment) Regulations 1992 (HS (G) 90)

Index

access equipment   52–4
accidents   100–1, 107, 109
action levels   93
appointed persons   113
Approved Codes of Practice   4, 25–31,
  35, 91, 93, 97, 112–16
asbestos   92–4, 101
Asbestos (Licensing) Regulations
  1983   94
Asbestos (Prohibitions) Regulations
  1992   94

Borehole Sites and Operations
  Regulations 1995   76
breach of contract   10
breach of statutory duty   11, 37
Building Regulations   73

Carriage of Dangerous Substances by
  Road Regulations 1996   76
CE mark   44
civil liability   10–12, 18–19, 37
Client   27–30, 32–3, 35, 37
Companies Directors Disqualification
  Act 1986   9
competent person   5, 58
Construction (Design and
  Management) Regulations 1994   24,
  53, 56, 114
Construction (General Provisions)
  Regulations 1961   56

Construction (Head Protection)
  Regulations 1989   86
Construction (Health and Welfare)
  Regulations 1966   56
Construction (Lifting Operations)
  Regulations 1961   44–5, 48
Construction (Working Places)
  Regulations 1966   56
Contractor   36–7, 71, 73, 101, 112
Control of Asbestos (Amendment)
  Regulations 1992   93
Control of Asbestos at Work
  Regulations 1987   87, 92
control levels   93
Control of Substances Hazardous to
  Health Regulations 1994   88–9
cranes   46–8, 51–3
Crown Court   8

dangerous occurrences   100–1
Dangerous Substances (Notification
  and Marking of Sites) Regulations
  1990   74
Designers   30–3, 71
diseases   101
display screen equipment   125–6,
  128–9

Electricity at Work Regulations
  1989   65
Employers' Liability (Compulsory
  Insurance) Act 1969   12

enforcement proceedings   1, 109
equipment guards   41
European Council Directive on work
   equipment   42
excavations   57, 59–62
explosives   97

Factories Act 1961   13, 21, 44, 48
fall prevention   51, 53, 58–60, 63
fire certificates   69–71, 74
Fire Precautions Act 1971   69–70
Fire Precautions (Factories, Offices,
   Shops and Railway Premises) Order
   1989   70–1
Fire Precautions (Workplace)
   Regulations 1997   72
Fire Safety and Safety of Places of
   Sport Act 1987   70
first-aid   111–16
first-aid box   112, 115–16
first-aider   113, 116
first-aid room   112, 114–16

Guidance Notes   4

head protection   80, 86–8
Health and Safety (Display Screen
   Equipment) Regulations 1992   65,
   125
Health and Safety File   27–9, 32–4
Health and Safety (First-Aid)
   Regulations 1981   111
Health and Safety Information for
   Employees (Modifications and
   Repeals) Regulations 1995   20
Health and Safety Information for
   Employees Regulations 1989   20
Health and Safety Plan   25, 27–9,
   32–5, 37
Health and Safety at Work etc. Act
   1974   2–3, 13–14
   Section 2   13–14, 18, 57, 61, 109
   Section 3   15–18, 81
   Section 4   18
   Section 37   9

Highly Flammable Liquids and
   Liquefied Petroleum Gases
   Regulations 1972   76, 98
hoists   51–2
HSE Form 2508   102–6
HSE Form 2530   47, 49
HSE Form 2531   47, 50

improvement notice   7
involuntary manslaughter   10
ionising radiation   94–7
Ionising Radiations (Outside Workers)
   Regulations 1993   96
Ionising Radiations Regulations 1985
   94

lifting equipment   44
Lifts Regulations 1997   54
lighting   42–3, 60, 63, 65–6, 72, 84,
   127

Magistrates' Court   8
Management of Health and Safety at
   Work Regulations 1992   4, 7, 19,
   21, 26, 35, 109, 114
manslaughter   9–10
manual handling   119–24
Manual Handling Operations
   Regulations 1992   65
maximum exposure level (MEL)
   91

negligence   11
noise   66–7, 87, 128
Noise at Work Regulations 1989   66,
   87

occupational exposure standard (OES)
   91
Occupiers' Liability Acts   18
Offices, Shops and Railway Premises
   Act 1963   13, 21

personal protective equipment   79–88,
   90–1, 93, 95

Personal Protective Equipment (EC Directive) Regulations 1992, as amended   67, 82
Personal Protective Equipment at Work Regulations 1992   81
Planning Supervisor   28–33
Principal Contractor   28–9, 32–7
prohibition notice   8
Provision and Use of Work Equipment Regulations 1992   21, 39–40, 65

R v. Associated Octel Co. Ltd   16–17
reasonably practicable   4, 9, 14–15, 17–18, 29–32, 34–6, 57–63, 66, 81, 90–2, 95
record keeping   102, 109, 114
Reporting of Injuries, Diseases and Dangerous Occurrences Regulations 1995   51, 99, 115

safety signs   129–31

Security (Claims and Payments) Regulations   107, 115
special premises   74–5
suitable persons   113
Supply of Machinery (Safety) (Amendment) Regulations 1994   43
Supply of Machinery (Safety) Regulations 1992   41–2, 66

Temporary or Mobile Construction Sites Directive   23, 56

vehicles   59–60, 76
ventilation   63, 72, 95

welfare   14, 21, 26, 55–6, 61–2, 64
Workplace (Health, Safety and Welfare) Regulations 1992   21, 55, 62
work equipment   21, 39–44, 65, 125–9